AF360704

Heat Exchangers for Waste Heat Recovery

Heat Exchangers for Waste Heat Recovery

Special Issue Editors

Petr Stehlik
Zdenek Jegla

MDPI • Basel • Beijing • Wuhan • Barcelona • Belgrade • Manchester • Tokyo • Cluj • Tianjin

Special Issue Editors

Petr Stehlik   Zdenek Jegla
Brno University of Technology   Brno University of Technology
Czech Republic   Czech Republic

Editorial Office
MDPI
St. Alban-Anlage 66
4052 Basel, Switzerland

This is a reprint of articles from the Special Issue published online in the open access journal *Energies* (ISSN 1996-1073) (available at: https://www.mdpi.com/journal/energies/special_issues/heat_exchangers_for_waste_heat_recovery).

For citation purposes, cite each article independently as indicated on the article page online and as indicated below:

LastName, A.A.; LastName, B.B.; LastName, C.C. Article Title. *Journal Name* **Year**, *Article Number*, Page Range.

ISBN 978-3-03936-475-6 (Hbk)
ISBN 978-3-03936-476-3 (PDF)

© 2020 by the authors. Articles in this book are Open Access and distributed under the Creative Commons Attribution (CC BY) license, which allows users to download, copy and build upon published articles, as long as the author and publisher are properly credited, which ensures maximum dissemination and a wider impact of our publications.

The book as a whole is distributed by MDPI under the terms and conditions of the Creative Commons license CC BY-NC-ND.

Contents

About the Special Issue Editors

Petr Stehlik (Prof. Dr.) is a Professor and Director of the Institute of Process Engineering of the Faculty of Mechanical Engineering at the Brno University of Technology. He is the Vice President of the Czech Society of Chemical Engineers and member of AIChE and ASME. He is executive editor or guest editor of several international journals; member or chairmen of scientific committees of international conferences; reviewer of PhD theses at reputable foreign universities; coordinator or contractor of international and national research projects; and the author and co-author of numerous publications and several patents. His research, development, and application activities involve heat transfer and its applications (heat exchangers, combustion systems, waste heat utilization); thermal treatment and energy utilization of waste (waste-to-energy); and energy savings and environmental protection.

Zdenek Jegla (Assoc. Prof. Dr.) is currently an Associate Professor and Senior Researcher in the field of Process Engineering at the Institute of Process Engineering of the Faculty of Mechanical Engineering (FME) at the Brno University of Technology (BUT) and a Head of the Heat Transfer Research Group at the NETME Research Centre FME BUT. He has many years of research and industrial experience, and he is author or co-author of numerous papers and contributions in international journals and conferences. His research and development as well as application activities are aimed at applied and enhanced heat transfer; waste heat recovery systems and equipment; process fired heaters, boilers, and heat exchangers; simulation, optimization, and CFD applications in the process and power industry; and process and equipment design and integration for energy savings and emissions reduction.

Preface to "Heat Exchangers for Waste Heat Recovery"

The economical and efficient recovery of waste heat produced by industrial processes (such as chemical and petrochemical processing, food, pharmaceutical, energetics) and processes and applications in the municipal sphere (such as waste incinerators, heating plants, laundries, hospitals, server rooms) are priorities and challenges. This Special Issue focuses on heat exchangers as key and essential equipment for the practical realization of these challenges. The purpose of this Special Issue is to outline the latest insights and innovative and/or enhanced solutions from the design, production, operation, and maintenance points of view of heat exchangers in different applications of effective waste heat utilization.

Petr Stehlik, Zdenek Jegla
Special Issue Editors

Article

Heat Transfer Characteristics of Heat Exchangers for Waste Heat Recovery from a Billet Casting Process

Ju O Kang and Sung Chul Kim *

School of Mechanical Engineering, Yeungnam University, 280 Daehak-ro, Gyeongsan-si, Gyeongbuk 712-749, Korea
* Correspondence: sungkim@ynu.ac.kr; Tel.: +82-53-810-2572; Fax: +82-53810-4627

Received: 21 June 2019; Accepted: 12 July 2019; Published: 15 July 2019

Abstract: The application of the thermoelectric generator (TEG) system to various industrial facilities has been explored to reduce greenhouse gas emissions and improve the efficiency of such industrial facilities. In this study, numerical analysis was conducted according to the types and geometry of heat exchangers and manufacture process conditions to recover waste heat from a billet casting process using the TEG system. The total heat absorption increased by up to 10.0% depending on the geometry of the heat exchanger. Under natural convection conditions, the total heat absorption increased by up to 45.5%. As the minimum temperature increased, the effective area increased by five times. When a copper heat exchanger of direct conduction type was used, the difference between the maximum and minimum temperatures was significantly reduced compared to when a stainless steel heat exchanger was used. This confirmed that the copper heat exchanger is more favorable for securing a uniform heat exchanger temperature. A prototype TEG system, including a thermosyphon heat exchanger, was installed and a maximum power of 8.0 W and power density of 740 W/m^2 was achieved at a hot side temperature of 130 °C. The results suggest the possibility of recovering waste heat from billet casting processes.

Keywords: waste heat recovery; cylindrical shape heat source; thermoelectric generator; radiative heat exchanger; numerical analysis; industrial experiment

---

1. Introduction

The global annual energy consumption is 474.1 PJ, of which 52% is discharged as waste heat in exhaust gas and effluents. Particularly in industrial areas, 22% of the total energy is being used annually but only half the amount is being used efficiently [1]. The low efficiency of industrial facilities affects greenhouse gas emissions, which are the main cause of global warming; therefore, the recovery of waste heat is expected to contribute to increasing economic efficiency and decreasing greenhouse gas emissions by improving the efficiency of facilities [2]. For this purpose, waste heat recovery technologies, including heat pumps, boilers, refrigeration cycles, and heat exchangers, have been developed. Among them, the application of the heat pipe and heat exchanger with a thermoelectric generator (TEG) and organic Rankine cycle (ORC) have been studied recently. The ORC system is a waste heat recovery technology that uses an organic refrigerant with low evaporation temperature to produce electric energy through the turbine. Peris et al. [3] installed an ORC system in a ceramic furnace and produced 21.8 kW of power with a maximum efficiency of 12.8% at 287 °C heat source temperature using exhaust gases. Ramirez et al. [4] installed an ORC system on a steel mill and obtained a 21.7% efficiency at a heat source of 529.6 °C. However, the ORC-based waste heat recovery system includes a heat exchanger, pump, and turbine, which limits its application due to its wide installation area. TEG is an eco-friendly energy conversion device that uses the Seebeck effect, in which current is generated from the temperature difference between high-temperature and low-temperature sections at both ends of two semiconductors. Although its current energy conversion efficiency is only 2–5%, a conversion efficiency of 15% or higher is expected to be attainable through the development of material technologies [5,6]. The application of the TEG system to internal combustion engines for waste heat recovery from high-temperature exhaust gas has already been investigated in the transport field. Wang et al. [7] fabricated a prototype TEG system to recover waste heat from the exhaust gas of hybrid vehicles and predicted potential improvement in fuel efficiency by up to 3.6% through experiments on power generation capacity and energy optimization. Moreover, Eddine et al. [8] performed an experiment using a TEG system for recovering waste heat from ship exhaust gas and predicted that maximum energy conversion efficiencies of 0.9% and 1.3% can be achieved with the use of Bi_2Te_3 at a maximum operating temperature of 250.0 °C and $Si_{80}Ge_{20}$ at 300.0 °C, respectively. In the industrial field, large amounts of energy can be generated using waste heat recovery systems even with low efficiency because there are various waste heat generation patterns and the scale of waste heat generation is large. For this reason, some studies have investigated the application of the TEG system for recovering industrial waste heat recently. Ebling et al. [9] installed a TEG system for the cooling process of steel forging products using a copper absorber plate to recover radiative waste heat, and produced 388 W of electric power at a heat source temperature of 1300 °C with an efficiency of 2.6%. Yazawa et al. [10] conducted research on the application of the TEG system to a glass melting furnace to recover waste heat. They analyzed various design parameters under the assumption that the TEG module of a direct conduction type was installed on the wall of the furnace, and predicted that a maximum electric power of 55.6 kW can be produced.

For the recovery of industrial waste heat, the TEG system has been applied to processes with high heat source temperatures (600.0 °C or higher), such as steel and glass manufacturing processes. In the case of non-ferrous metal industries, such as aluminum, brass, and copper; however, the temperature of the heat source at which waste heat is generated is in the mid-range, i.e., 300.0–600.0 °C, and thus the power generated by TEG is significantly reduced. Therefore, for processes with mid-range heat source temperature, the energy conversion efficiency of the TEG module as well as the heat transfer efficiency from the heat source to the TEG module must be considered in order to produce a large amount of energy. In addition, it is necessary to conduct research on the geometry and types of heat exchangers as important design parameters of the TEG system according to the waste heat generation patterns. In particular, for radiative energy, plate-type heat exchangers have been mostly used in the TEG system for heat sources located on horizontal surfaces, such as conveyor belts [9,11]. In actual processes, however, waste heat is generated from more complex geometry and plate-type heat exchangers cannot recover heat efficiently. Thus, some studies have been conducted to recover waste heat using heat pipes. A heat pipe consists of an evaporation unit, an insulation unit, and a condensation unit. The heat exchanger transfers heat through phase change via evaporation and condensation of the internal working fluid. The applicable temperature range of the heat source is wide and the heat transfer efficiency is high. For the application of heat pipes to industrial facilities, various geometries and performance parameters have been investigated through experiments and numerical analysis [12–14]. Regarding studies on the use of heat pipes in the TEG system, Huang et al. [15] confirmed a maximum output of 6.25 W in a TEG module using a loop heat pipe, and Wu et al. [16] showed that the use of heat pipes increased the TEG power generation efficiency by up to 50% in a comparative experiment using heat pipes and plates as heat spreaders. Heat pipe pyroelectric system with energy harvesting technology is similar to heat pipe TEG system. However, pyroelectric devices are more suitable where the temperature of the heat source changes over time [17,18]. The TEG system seems to be suitable because the heat source of this study is less temperature change with time.

This study attempted to minimize heat loss in a brass billet cooling process by applying a heat exchanger with the geometry of the area surrounding the billet such that heat could be more efficiently absorbed than plate-type heat exchangers. The TEG system was applied for the first time to the billet cooling process with a cylindrical heat source that generates waste heat of 600.0 °C or less. For the application of the heat exchanger, the difference in heat absorption and surface temperature distribution were analyzed according to the geometry of the heat exchanger and process conditions through numerical analysis. In addition, the reliability of the numerical analysis of the target heat source was verified through experiments. A prototype TEG system with a thermosyphon heat pipe was installed in the billet casting cooling process. Experiments on the output power and measurement of the temperatures of the high-temperature and low-temperature sections of the heat pipe and TEG module were performed.

2. Experimental Setup and Numerical Analysis

2.1. Experimental Setup

In this study, the brass billet cooling process in Gwangmyeong-si, Korea, was set as the target of the TEG system for waste heat recovery. The side exposed to the ambient after the billet continuous casting process was set as the target position, as shown in Figure 1. The billet diameter was 240 mm and the billet length exposed to the outside was 600 mm. An air injector was installed at the outlet of the casting machine to cool the billet surface, and the billet entered the water injector through an air cooling section. Heat loss from the billet at the waste heat recovery position was caused by natural convection due to the high-temperature surface, forced convection due to the air injector, and radiation around the billet. Regarding the heat exchanger for waste heat recovery from the billet, a non-contact heat exchanger is required because the billet moves after casting. As the flow around the billet is also not constant due to the air injector, the waste heat is recovered through radiative heat exchange.

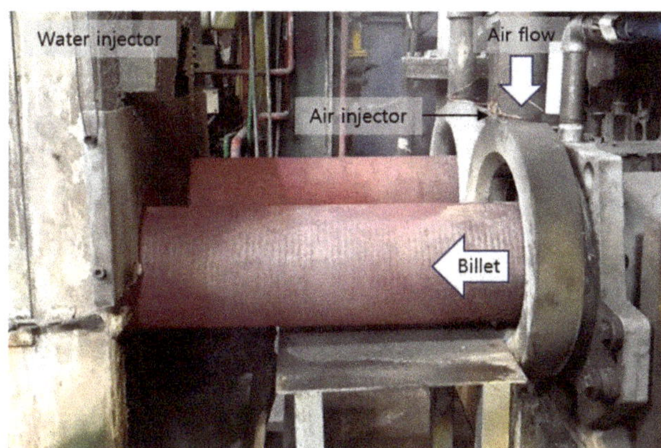

Figure 1. Brass billet continuous casting process.

A temperature measurement experiment was performed to verify the analysis. Figure 2 shows the schematic diagram of the experimental apparatus. An air injector was installed at the outlet of the casting machine to inject air towards the billet and a centrifugal fan capable of providing a volumetric flow rate of up to 63 m^3/min was connected. A SUS 316 absorber plate was installed at 20 mm from the billet to measure the radiative heat temperature between the water injector and the air injector. The length, height, and thickness of the plate were 500, 290, and 5 mm, respectively. For temperature measurement, three thermocouples were attached to the center of the specimen at 105 mm height intervals. An additional thermocouple was attached to measure the temperature distribution on the billet surface. As the billet was cast and moved, the thermocouples also moved and the surface temperatures of the billet were measured at a distance from the outlet of the casting machine. Table 1 shows the specifications of the experimental apparatus. In the experiment, the temperature of the specimen was judged to reach a steady state when temperature changes in the three thermocouples attached to the absorber plate were ±3.0 °C or less considering the errors of the thermocouples and the data logger. The specimen reached a steady state in approximately 30 min. The average temperatures at each position were calculated by collecting temperature data for more than 10 min and they were used as result data.

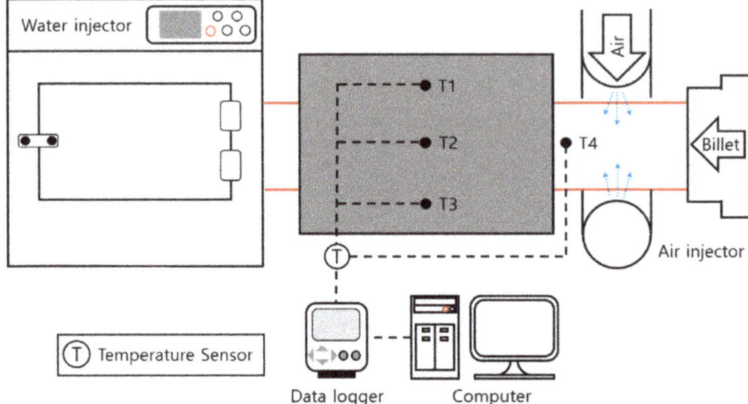

Figure 2. Schematic diagram of the experimental apparatus for analysis verification.

Table 1. Specifications of the experimental apparatus for analysis verification.

Sensor and Measure Instrument	Specification
Thermocouple (Omega, K-type)	±0.75% (−270.0–1372.0 °C) 3 points absorber plate, 1 point billet
Data logger (Kimo)	4 channels (−200.0–1372.0 °C)
Absorber plate (SUS 316)	400 (W) × 290 (H) × 5 (t)
Blower (Kigeonsa)	Centrifugal fan Max volume flow rate: 63 m^3/min

Figure 3a shows the schematic diagram of the radiative waste heat recovery TEG system for the experiment on output power. Radiative waste heat from the billet was recovered using the SUS 316 heat pipe. The working fluid of the heat pipe evaporated on the surface facing the billet and condensed on the top plate with the TEG module. Three K-type thermocouples were attached to measure the temperatures of the heat pipe condensation section as well as the high-temperature and low-temperature sections of the TEG module. The charging controller of the maximum power point tracking (MPPT) type was used to charge the battery with electrical energy generated by the TEG module. The battery charging power from the MPPT controller and the temperatures measured from the heat pipe and TEG module were transferred to the monitoring system through the main controller to enable real-time data collection and verification. Bi_2Te_3 was used in the installed TEG module. This material affords a maximum conversion efficiency of 3.4% when the maximum operating temperature of the high-temperature section is 300.0 °C and the temperature difference between both ends is 250 °C. Three modules with width, depth, and height of 60, 60, and 3.7 mm, respectively, were used. Figure 3b is a schematic diagram of the heat pipe composed of a single wick. The radius of the heat absorption plate of the wick, R_a, is 130 mm, the radius of the heat insulation plate, R_i, is 160 mm, the height of the wick portion, H, is 160 mm, and the total axial length is 500 mm. The filling ratio of the internal working fluid was set to 30%. In the heat pipe, DOWTHERM-A, a mixture of $C_{12}H_{10}$ and $C_{12}H_{10}O$, was used as the working fluid, with the operating temperature ranging from 15.0 to 400.0 °C. Table 2 shows all the specifications of the TEG system.

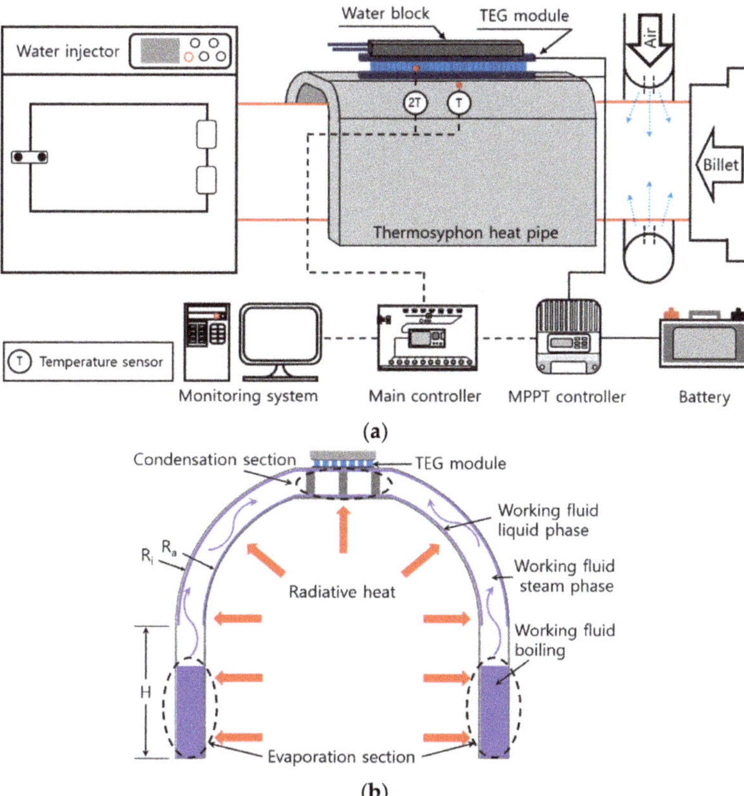

Figure 3. Schematic diagrams of the thermoelectric generator (TEG) system and heat pipe (**a**) TEG system; (**b**) heat pipe.

Table 2. Specifications of the TEG system.

Component	Specification
TEG modules	Bi_2Te_3, 3.4% at dT: 250 °C 60 (W) × 60 (L) × 3.7 (H) 3 ea
Heat pipe (SUS 316)	Type: Thermosyphon Working fluid: DOWTHERM-A Operating Temperature: (15.0–400.0 °C)
MPPT controller	Max input power: 260 W-12 V, 520 W-24 V
Main controller (Espressif)	Apply 10 temperature sensors

2.2. Numerical Analysis Model and Boundary Conditions

The heat exchanger for waste heat recovery was modeled in three geometrical shapes, as shown in Figure 4. The first model is the n-type shown in Figure 4a. It targeted at heat transfer to the TEG module installed at the top using the heat pipe. Flat plates were applied to the sides such that the fixing jig could be installed at the bottom. The second model is the o-type shown in Figure 4b. It is a model for comparing differences in heat absorption and temperature distribution according to the heat exchanger geometry. The flat plates on the side of the n-type were bent towards the billet such that radiative waste heat could be absorbed effectively. The third model is the d-type shown in Figure 4c. It targeted at heat transfer through direct conduction after the recovery of radiative waste heat. The TEG module can be installed both on the side and at the top, unlike the n-type and o-type. SUS 316 was used for the n-type and o-type considering the mechanical durability of the heat pipe, and they had the same heat exchange area of 0.4 m^2. Copper was used for the d-type for effective heat conduction through direct conduction. Its heat exchange area is the same as those of the n-type and o-type. Table 3 shows the specifications of all heat exchanger models.

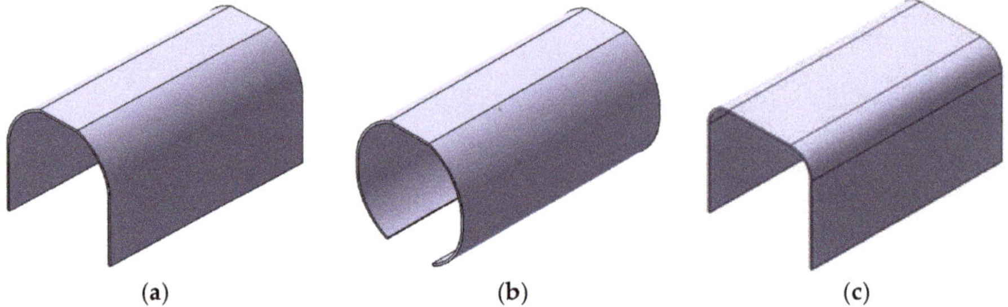

Figure 4. Geometry of the waste heat exchanger absorber plate. (**a**) n-Type; (**b**) o-Type; (**c**) d-Type.

Table 3. Specifications of heat exchanger model.

Type	n-Type	o-Type	d-Type
Material	SUS 316		Copper
Thermal conductivity (W/m·K)	14.8		65.0
Length (mm)		500.0	
Thickness (mm)		5.0	
Heat transfer area (m^2)		0.4	
Heat exchanger type	Heat pipe		Direct conduction

ANSYS Fluent (ver. 19.2), a commercial software program, was used for heat flow analysis in this study. Table 4 shows the analysis method. The analysis was conducted on the domain, including the billet, air injector, and heat exchanger. Air was used as an incompressible ideal gas under steady-state, incompressible flow, and turbulent conditions. Moreover, the convergence of the residual was increased using the coupled algorithm and pseudo transient approach. The realizable k-epsilon model was used to simulate the turbulent flow around the billet and heat exchanger caused by the air injector. The scalable wall function was used to compensate for the grid quality degradation occurring from the narrow space between the billet and heat exchanger. The pressure was set as the body force weighted scheme considering the occurrence of the natural convection effect on the surface of the billet. The radiative heat exchange between the billet and heat exchanger is the main heat exchange method in this analysis and the surface to surface (S2S) model was used as an analytical model. The S2S model calculates radiative heat exchange using the view factor and assumes the heat exchange medium as a gray body and a diffusive surface. The absorption coefficient and the scattering coefficient could be neglected in the model, and the model was suitable for the analysis because there was no permeable medium in the analytical domain of this study [19]. For a somewhat larger scale with geometry similar to that of this study, Mirhosseini et al. [20] conducted 2D numerical analysis on the radiative heat exchange of an arc absorber around a cylindrical cement kiln using an S2S model. Table 5 shows the boundary conditions used for heat flow analysis. Emissivity, a surface property, was set to 0.7 for the brass billet, 0.4 for SUS 316, and 0.5 for copper [21–23]. Figure 5 shows the results of measuring the billet surface temperature. Considering the process, the average temperature of 530 °C at the position where the heat exchanger can be installed was set as the surface temperature of the billet. To compare the heat absorption and temperature distribution under forced convection and natural convection conditions, blower flow rates of 0 and 0.18 kg/s were added to each geometrical shape.

Table 4. Analysis method.

Component	Model and Boundary Conditions
State	Steady state, Incompressible ideal gas
Algorithm	Coupled, Pseudo transient
Pressure	Body force weighted
Turbulence	Realizable k-epsilon
Wall function	Scalable wall function
Radiation	S2S

Table 5. Analysis of boundary conditions.

Component	Boundary Condition
Emissivity	ε_b: 0.7, ε_{ss}: 0.4, ε_{cp}: 0.5
Billet temperature (°C)	530.0
Mass flow rate (kg/s)	Forced convection: 0.2 Natural convection: 0

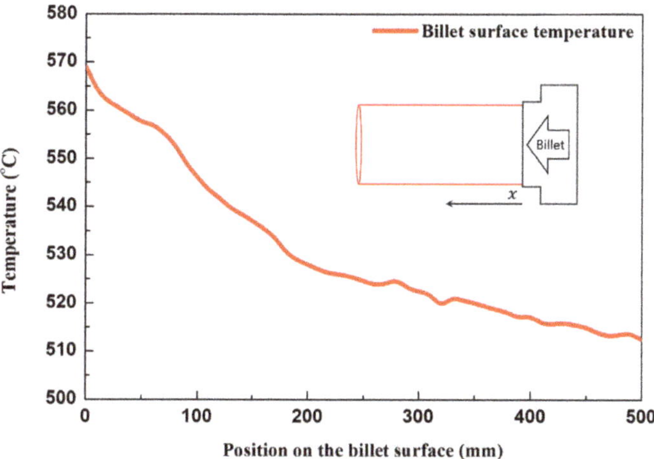

Figure 5. Temperature distribution on the billet surface.

3. Results and Discussion

Approximately 15 million grids were used to ensure grid independence of the analytical model. Convergence was determined when the residual value of energy dropped below 10^{-6} and the remaining residual value dropped below 10^{-4}. The reliability of the analytical model was confirmed through the experiment for analysis verification and the maximum error of 2.4%. Table 6 shows temperatures and errors at each measuring point obtained from the numerical analysis and the experiment.

Table 6. Temperature difference between experimental and numerical analysis.

Measuring Point	Experimental Result (°C)	Numerical Result (°C)	Error (%)
T1	135.6	138.9	2.4
T2	169.4	171.3	1.1
T3	151.9	154.2	1.5

Figure 6 shows the temperature distribution according to the geometry of the SUS 316 heat exchanger. Under forced convection conditions, the high-temperature section on the heat exchanger surface was formed at the front for all geometrical shapes. Figure 6a shows the surface temperature distribution of the n-type under the forced convection condition. The maximum temperature was 272.5 °C, whereas the minimum temperature was 75.5 °C. Figure 6c shows the surface temperature distribution of the o-type under the same condition. The maximum temperature was 274.1 °C, whereas the minimum temperature was 94.1 °C. The difference in the maximum temperature between the n-type and o-type was small (1.6 °C), but the difference in the minimum temperature was relatively large (18.6 °C) with the o-type exhibiting higher minimum temperature. The asymmetric temperature distribution of the symmetric heat exchanger could be predicted because the position of the injection hole on the air injector is not symmetrical. Figure 6b and d show the heat exchanger surface temperature distributions of the n-type and o-type, respectively, under the natural convection condition. The maximum temperature was 348.4 °C for the n-type and 346.4 °C for the o-type, showing an increase of up to 75.9 °C compared to the maximum temperatures obtained under the forced convection condition. This can be attributed to the increase in the temperature of the heat exchanger with the weakening of the effect of forced convection, reducing heat loss on the surface. Moreover, the minimum temperature was 198.1 °C for the n-type and 286.3 °C for the o-type, showing an increase of up to 192.2 °C. The temperature difference between the n-type and o-type was obvious under the natural convection condition. Figure 6e shows the surface temperature distribution of the d-type under the forced convection condition. The maximum temperature was 221.6 °C, whereas the minimum temperature was 148.4 °C. Figure 6f shows the surface temperature distribution of the d-type under the natural convection condition. The maximum and minimum temperatures were 280.2 °C and 243.0 °C, respectively. While the difference between the maximum and minimum temperatures was up to 197 °C for the n-type and o-type, that of the d-type was up to 73.2 °C, which is significantly low. The difference between the maximum and minimum temperatures of the d-type appeared to decrease because of the high thermal conductivity of copper. For the d-type, a uniform temperature distribution is required for the heat exchanger because heat is transferred to the TEG module through direct conduction. The analysis results showed that the use of the copper heat exchanger is appropriate because it exhibits a relatively uniform temperature distribution.

As shown in Figure 6g,h, a convection layer was formed between the billet and the heat exchanger. The high-temperature section was concentrated at the front because the cooling air that entered the rear end of the heat exchanger close to the air injector was slowly heated as it passed between the billet and the heat exchanger. Figure 7 shows the temperature distribution along the heat exchanger circumference. The maximum temperature was not significantly different depending on the geometry, but the temperature significantly increased in the o-type due to the bent sides.

Although stainless steel was selected considering the mechanical durability of the heat pipe, the temperature difference between the high-temperature and low-temperature sections was large due to its low thermal conductivity. To determine the area where the working fluid can evaporate and heat absorbed when heat pipes are used, the effective area and waste heat absorption were analyzed according to the geometrical and flow conditions for the basic design of the heat exchanger. The effective area was defined as the area where the temperature of the heat absorbing section was 250 °C or higher considering that the maximum operating temperature of the high-temperature section of the TEG module was 300 °C. The effective area and waste heat absorption, according to the geometry and flow conditions, are shown in Figure 8. Under the forced convection condition, a total of 1.0 kW waste heat was absorbed by the heat exchanger for the n-type. For the o-type under the same condition, the radiative heat energy absorbed from the billet increased due to the bent sides and 1.1 kW waste heat was absorbed. Under the natural convection condition, the energy loss attributable to convection decreased and thus the n-type and o-type absorbed 1.5 kW and 1.6 kW waste heat, respectively. The effective area did not significantly differ depending on the geometry, and it significantly increased under the natural convection condition. Thus, most of the area exhibited 250 °C or higher for both the n-type and o-type, indicating effective heat transfer. The numerical analysis results confirmed that the geometry contributed to 10% improvement of waste heat absorption but it did not significantly affect the maximum temperature and effective area. Under the natural convection condition, however, waste heat absorption was increased by up to 45.5% and the effective area was increased up to five times.

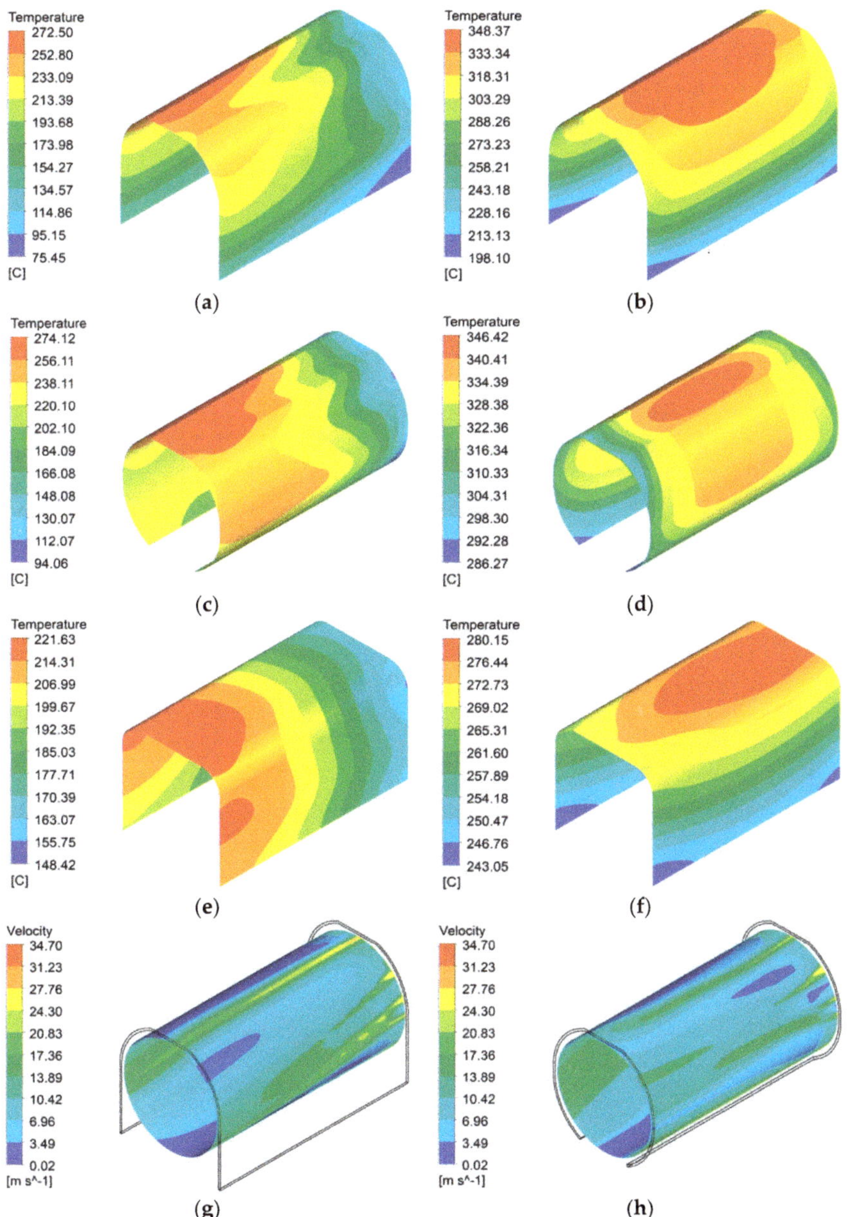

Figure 6. Temperature distribution with different heat exchanger shapes and air flow and velocity distribution between the billet and the heat exchanger (**a**) n-Type forced convection; (**b**) n-Type natural convection; (**c**) o-Type forced convection; (**d**) o-Type natural convection; (**e**) d-Type forced convection; (**f**) d-Type natural convection; (**g**) n-Type forced convection; (**h**) o-Type forced convection.

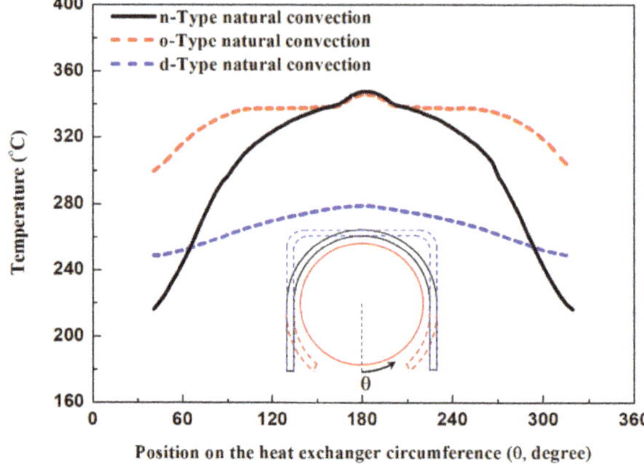

Figure 7. Surface temperature distribution along the heat exchanger circumference.

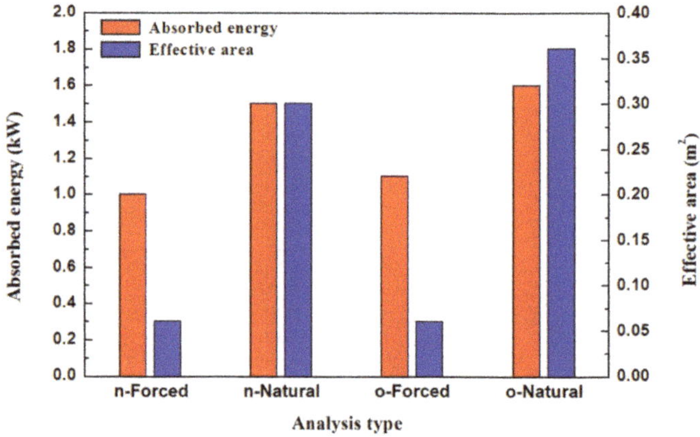

Figure 8. Heat absorption with different heat exchanger shapes and air flow.

The TEG system with an n-type prototype thermosyphon was tested under the forced convection condition. Table 7 shows the temperatures and output power of the TEG system in a steady state. In the actual experiment, the temperature of the condensation section of the heat exchanger was 263.3 °C, meeting the proper operating temperature of the TEG module. In the numerical analysis, the temperature of the high-temperature section at the top of the n-type was similar to that at the top of the thermosyphon with an error of 3.5%. The temperature of the high-temperature section of the TEG module was 130 °C, which was also appropriate as it was in the operating temperature range. The three TEG modules provided an output power of 8.0 W when the temperature of the low-temperature section was 49.8 °C and the temperature difference between both ends was 80.2 °C. The temperature drop that occurred at the condensation section of the heat exchanger and the high-temperature section of the TEG module; however, was a major obstacle to the system output. This temperature drop can be attributed to the insufficient waste heat recovery and low thermal conductivity of the TEG system. Regarding the evaporation section, as confirmed by the numerical analysis, the evaporation of the working fluid was not sufficient because the thermosyphon evaporation section could not secure sufficient temperature and heat absorption, thereby, degrading the performance of the heat exchanger itself. As confirmed through the numerical analysis, heat absorption was increased by 50.0% under the natural convection condition. Therefore, if the air injector is removed from the TEG system as a means of securing output and an experiment is performed under the natural convection condition, the output of the TEG system is expected to increase. The heat exchanger output can also be increased by increasing the emissivity of the thermosyphon through surface treatment, such as surface blackening using carbon coating.

Table 7. Temperatures and output power of the TEG system.

Type	Condenser Temperature (°C)	TEG Hot Side Temperature (°C)	TEG Cold Side Temperature (°C)	Output (W)
Experimental	263.3	130	49.8	8.0

4. Conclusions

In this study, the heat absorption and temperature distribution of a TEG system for recovering waste heat from a brass billet casting process were analyzed according to the heat exchanger type, geometry, and process condition using numerical analysis. Heat absorption could be increased by up to 10.0% depending on the geometrical variables, but the maximum temperature did not significantly change. Moreover, when the forced convection condition was changed to the natural convection condition, of which the blower flow rate was 0 kg/s, heat absorption could be increased by up to 45.5%, the maximum temperature was also increased by 75.9 °C, and the effective area increased five times, thereby, exhibiting more effective heat transfer. The heat exchanger of a direct conduction type also met the proper operating temperature of the TEG module and exhibited a uniform temperature distribution, thereby, suggesting its feasibility for practical application. Moreover, the experiment on the output power of the TEG system with the n-type thermosyphon under the forced convection condition confirmed the output power of 8.0 W.

The results of this study confirmed the possibility of recovering waste heat from industrial facilities. The application of the TEG system is expected to improve the efficiency of facilities and contribute to the selection of heat exchangers for the TEG system according to the type of waste heat generated. The system outputted only 8 W during the actual process because the heat transfer between the heat exchanger and TEG module has not been optimized yet. In the future, for the optimal design of heat exchangers and improvement of the output power, various types of heat exchangers and heat exchangers with coated surfaces for improving heat transfer performance will be applied to the TEG system and experiments as well as numerical analysis on power generation performance and economic assessment of its applicability will be conducted.

Author Contributions: J.O.K. performed the experimental/numerical analysis and drafted the manuscript. S.C.K. organized the overall evaluation and reviewed the manuscript.

Funding: This work was conducted as a part of the Energy Technology Development project sponsored by the Ministry of Trade, Industry and Energy (No. 20172010000760).

Acknowledgments: The support received from the Livingcare company is greatly appreciated.

Conflicts of Interest: The authors declare no conflict of interest.

Nomenclatures

H	Height
R	Radius
t	Thickness (mm)
ε	Emissivity

Subscripts

a	Absorption plate
b	Billet
cp	Copper
i	Insulation plate
ss	Stainless steel

References

1. Forman, C.; Muritala, I.K.; Pardemann, R.; Meyer, B. Estimating the global waste heat potential. *Renew. Sustain. Energy Rev.* **2016**, *57*, 1568–1579. [CrossRef]
2. Firth, A.; Zhang, B.; Yang, A. Quantification of global waste heat and its environmental effects. *Appl. Energy* **2019**, *235*, 1314–1334. [CrossRef]
3. Peris, B.; Navarro-Esbri, J.; Moles, F.; Mota-Babiloni, A. Experimental study of an ORC (organic Rankine cycle) for low grade waste heat recovery in a ceramic industry. *Energy* **2015**, *85*, 534–542. [CrossRef]
4. Ramirez, M.; Epelde, M.; Gomez de Arteche, M.; Panizza, A.; Hammerschmid, A.; Baresi, M.; Monti, N. Performance evaluation of an ORC unit integrated to a waste heat recovery system in a steel mill. *Energy Procedia* **2017**, *129*, 535–542. [CrossRef]
5. Huang, F.; Zheng, J.; Baleynaud, J.M.; Lu, J. Heat recovery potentials and technologies in industrial zones. *J. Energy Inst.* **2017**, *90*, 951–961. [CrossRef]
6. Jouhara, H.; Khordehgah, N.; Almahmoud, S.; Delpech, B.; Chauhan, A.; Tassou, S.A. Waste heat recovery technologies and applications. *Therm. Sci. Eng. Prog.* **2018**, *6*, 268–289. [CrossRef]
7. Wang, R.; Yu, W.; Meng, X. Performance investigation and energy optimization of a thermoelectric generator for a mild hybrid vehicle. *Energy* **2018**, *162*, 1016–1028. [CrossRef]
8. Eddine, A.N.; Chalet, D.; Faure, X.; Aixala, L.; Chesse, P. Optimization and characterization of a thermoelectric generator prototype for marine engine application. *Energy* **2018**, *143*, 682–695. [CrossRef]
9. Ebling, D.G.; Krumm, A.; Pfeiffelmann, B.; Gottschald, J.; Bruchmann, J.; Benim, A.C.; Adam, M.; Labs, R.; Herbertz, R.R.; Stunz, A. Development of a system for thermoelectric heat recovery from stationary indusrial processes. *J. Electron. Mater.* **2016**, *45*, 3433–3439. [CrossRef]
10. Yazawa, K.; Shakouri, A.; Hendricks, T.J. Thermoelectric heat recovery from glass melt processes. *Energy* **2017**, *118*, 1035–1043. [CrossRef]
11. Kuroki, T.; Kabeya, K.; Makino, K.; Kajihara, T.; Kaibe, H.; Hachiuma, H.; Matsuno, H.; Fujibayashi, A. Thermoelectric generation using waste heat in steel works. *J. Electron. Mater.* **2014**, *43*, 2405–2410. [CrossRef]
12. Shabgard, H.; Allen, M.J.; Sharifi, N.; Benn, S.P.; Faghri, A.; Bergman, T.L. Heat pipe heat exchangers and heat sinks: Opportunities, challenges, applications, analysis, and state of the art. *Int. J. Heat Mass Transf.* **2015**, *89*, 138–158. [CrossRef]
13. Jouhara, H.; Almahmoud, S.; Chauhan, A.; Delpech, B.; Bianchi, G.; Tassou, S.A.; Llera, R.; Lago, F.; Arribas, J.J. Experimental and theoretical investigation of a flat heat pipe heat exchanger for waste heat recovery in the steel industry. *Energy* **2017**, *141*, 1928–1939. [CrossRef]
14. Delpech, B.; Axcell, B.; Jouhara, H. Experimental investigation of a radiative heat pipe for waste heat recovery in a ceramics kiln. *Energy* **2019**, *170*, 636–651. [CrossRef]
15. Huang, B.J.; Hsu, P.C.; Tsai, R.J.; Hussain, M.M. A thermoelectric generator using loop heat pipe and design match for maximum-power generation. *Appl. Therm. Eng.* **2015**, *91*, 1082–1091. [CrossRef]
16. Wu, S.; Ding, Y.; Zhang, C.; Xu, D. Improving the performance of a thermoelectric power system using a flat-plate heat pipe. *Chin. J. Chem. Eng.* **2019**, *27*, 44–53. [CrossRef]
17. Nguyen, H.; Navid, A.; Pilon, L. Pyroelectric energy converter using co-polymer P(VDF-TrFE) and Olsen cycle for waste heat energy harvesting. *Appl. Therm. Eng.* **2010**, *30*, 2127–2137. [CrossRef]
18. Poprawski, W.; Gnutek, Z.; Radojewski, J.; Poprawski, R. Pyroelectric and dielectric energy conversion – A new view of the old problem. *Appl. Therm. Eng.* **2015**, *90*, 858–868. [CrossRef]
19. FLUENT Inc. *Fluent 6.3 User's Guide*; FLUENT Inc.: Havover, NH, USA, 2006.
20. Mirhosseini, M.; Rezaniakolaei, A.; Rosendahl, L. Numerical study on heat transfer to an arc absorber designed for a waste heat recovery system around a cement kiln. *Energies* **2018**, *11*, 671. [CrossRef]
21. Shi, D.; Liu, Q.; Zhu, Z.; Sun, J.; Wang, B. Experimental study of the relationships between the spectral emissivity of brass and the temperature in the oxidizing environment. *Infrared Phys. Technol.* **2014**, *64*, 119–124. [CrossRef]

22. Nam, H.Y.; Lee, G.Y.; Kim, J.M.; Choi, S.K.; Park, J.H.; Choi, I.G. Experimental Study on the Emissivity of Stainless Steel. In Proceedings of the Korean Nuclear Society Spring Meeting, Cheju, Korea, 24–25 May 2001; Available online: https://inis.iaea.org/search/searchsinglerecord.aspx?recordsFor=SingleRecord&RN=33040336 (accessed on 20 May 2019).
23. Incropera, F.P.; Dewitt, D.P. *Foundations of Heat Transfer*, 6th ed.; John Wiley & Sons: New York, NY, USA, 2013.

© 2019 by the authors. Licensee MDPI, Basel, Switzerland. This article is an open access article distributed under the terms and conditions of the Creative Commons Attribution (CC BY) license (http://creativecommons.org/licenses/by/4.0/).

Article

Modelling and Evaluation of Waste Heat Recovery Systems in the Case of a Heavy-Duty Diesel Engine

Amin Mahmoudzadeh Andwari [1,2,*], Apostolos Pesyridis [1], Vahid Esfahanian [2], Ali Salavati-Zadeh [3] and Alireza Hajialimohammadi [4]

[1] Centre for Advanced Powertrain and Fuels Research (CAPF), Department of Mechanical, Aerospace and Civil Engineering, Brunel University London, London UB8 3PH, UK; a.pesyridis@brunel.ac.uk
[2] Vehicle, Fuel and Environment Research Institute (VFERI), School of Mechanical Engineering, College of Engineering, University of Tehran, Tehran 1439956191, Iran; evahid@ut.ac.ir
[3] Niroo Research Institute (NRI), Tehran 1468613113, Iran; asalavatizadeh@nri.ac.ir
[4] Mechanical Engineering Department, Semnan University, Semnan 3513119111, Iran; ahajiali@semnan.ac.ir
* Correspondence: amin.mahmoudzadehandwari@brunel.ac.uk; Tel.: +44-(0)-1895-267901

Received: 3 March 2019; Accepted: 26 March 2019; Published: 11 April 2019

Abstract: In the present study, the effects of Organic Rankine Cycle (ORC) and turbo-compound (T/C) system integration on a heavy-duty diesel engine (HDDE) is investigated. An inline six-cylinder turbocharged 11.5 liter compression ignition (CI) engine employing two waste heat recovery (WHR) strategies is modelled, simulated, and analyzed through a 1-D engine code called GT-Power. The WHR systems are evaluated by their ability to utilize the exhaust excess energy at the downstream of the primary turbocharger turbine, resulting in brake specific fuel consumption (BSFC) reduction. This excess energy is dependent on the mass flow rate and the temperature of engine exhaust gas. However, this energy varies with engine operational conditions, such as speed, load, etc. Therefore, the investigation is carried out at six engine major operating conditions consisting engine idling, minimum BFSC, part load, maximum torque, maximum power, and maximum exhaust flow rate. The results for the ORC and T/C systems indicated a 4.8% and 2.3% total average reduction in BSFC and also maximum thermal efficiencies of 8% and 10%, respectively. Unlike the ORC system, the T/C system was modelled as a secondary turbine arrangement, instead of an independent unit. This in turn deteriorated BSFC by 5.5%, mostly during low speed operation, due to the increased exhaust backpressure. It was further concluded that the T/C system performed superiorly to the ORC counterpart during top end engine speeds, however. The ORC presented a balanced and consistent operation across the engines speed and load range.

Keywords: waste heat recovery; Organic Rankine Cycle; turbo-compound; brake specific fuel consumption; engine thermal efficiency

---

1. Introduction

In recent years, the growing worldwide population and industrial development has seen an equally increasing demand in energy. The internal combustion engine (ICE) has, by far, grown to be the most popular means of transport since the second half of the 20th century. Unfortunately, a typical ICE will only manage to convert approximately 30–35% of the total provided chemical energy into effective mechanical work, as illustrated in Figure 1 [1–6].

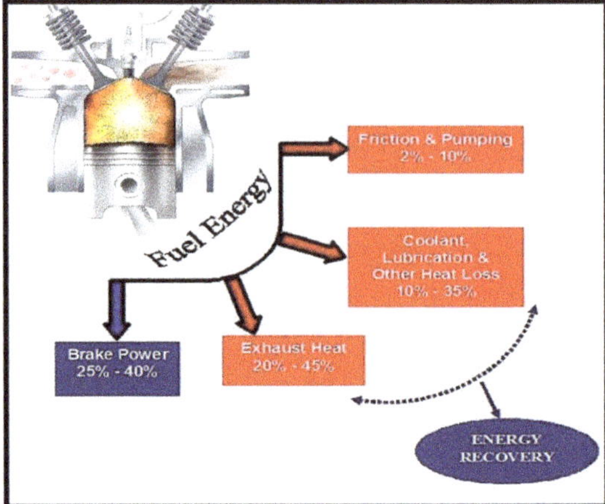

Figure 1. Variation total fuel energy content in ICE.

As with any system that produces work in real life, it is very challenging to achieve an adiabatic process during which thermal expansion occurs. Even in present ICEs, approximately 60–70% of the energy discharged by the fuel is wasted predominately in the form of heat [7–13]. During the last two decades, typical engine specific power output ratios are in the region of 1.5. At the same time, emission levels have reached a factor of 10. This is mainly due to the restrictions imposed by EURO, forcing technology to progress in the production of cleaner and more efficient ICE [14–20]. Even a conventional turbocharger only takes advantage of a portion of the exhaust gas energy in the shape of kinetics and pressure, which constitutes a fraction of the energy losses in a naturally aspirated engine. The biggest percentage is heat transfer and exhaust gas enthalpy dissipation, which is accountable for about 50–85% of the outstanding low heating values of the utilized fuel [21–25]. As a result, presently, waste heat recovery (WHR) has been the primary concentration point of research and development departments of engine manufacturers. This is due to the considerable potential energy amount that can be recovered in the form of heat [26–28]. Modern WHR systems amongst others include the following:

- Mechanical turbo-compounding
- Electrical turbo-compounding
- Thermoelectric generators (TEG)
- Steam Rankine cycle
- Organic Rankine cycle (ORC)
- Brayton thermodynamic cycle

All of the above can recover a segment of the exhaust gas energy and can subsequently enhance the engines' thermal efficiency. They all operate on relatively similar thermodynamic principles, but not all of them perform in the same fashion. The beginning of research and development on the feasibility of the Rankine cycle system as a WHR method dates back to the 1970's [29–32]. The fundamental orientation of the studies has been the thermal optimization of heavy duty diesel engines (HDDE) due to their high thermal efficiency potential, ranging, by todays' standards, between 40–45% [33–36]. This makes HDDE the most favorable candidates, as a 10–15% improvement in fuel efficiency is not uncommon for ORC applications [37–40]. The high thermal efficiency figures have also influenced industries to make use of HDDE, with power outputs of up to 600 kW for on and off-highway commercial vehicles. Typical engine displacements vary between 6–12 liter with multiple cylinder numbers and configurations, while the latest era of HDDE, utilizing high-pressure common rail direct injection systems, reach pressures of over 2500 bar. Part of the reason for this high efficiency is the use of forced induction. Forced induction constitutes a method of increasing engine power output that dates back to the late nineteenth and early twentieth century [41–45].

Supercharging an engine can be done both by either mechanical and/or chemical means. Power output is directly proportional to the mass of fuel and air burned inside the engines' cylinders. To ncrease fuel mass delivery, one must first increase air mass intake to avoid continuous combustion of rich mixture. Additional air supplying components that are gear and belt driven are normally called superchargers or blowers. Turbocharging is when the supercharger is driven by the engines' hot exhaust gases. One of the reasons forced induction became so popular over the past decades, other than the significant performance gains, is because it was a clever way of benefiting from the energy of burned exhaust gas residue that would otherwise be wasted. As a result, the automotive industry perceived it to be an acceptable mix of cost, performance, fuel economy, and reliability [40,46–48]. In turbocharged engines, heat transfer is a very complex entity, which greatly affects turbocharger performance, efficiency, and selection. As exhaust gas flows through the turbine, the turbine housing absorbs a sizable percentage of the total enthalpy by forced convection, due to the temperature difference between the walls and exhaust gases. This heat is then lost to the environment by means of radiation. Heat transfers by forced convection are also evident on the turbine wheel blades, shaft, and, subsequently, on the lubricant because of the exhaust gas expansion [49–54]. On the other hand, the compressor side acts as a heat sink and is subject to heat conduction derived through the bearing and/or turbine housings as well as the engine itself. This heat flux is inbound and may affect the temperature and pressure of the inducted air at low rotating speeds and compression ratios [55,56]. It was discovered that despite past research on WHR using primarily ORC and T/C systems either of mechanical or electrical nature, there is limited information on the comparison of the two systems, which are assessed in this research [1,36,57–59]. This study aims to identify which of the two adopted WHR methods (ORC and T/C) performs in a better manner in terms of improving engine thermal efficiency and BSFC by means of 1-D engine simulation software.

1.1. Waste Heat Recovery

Almost 60% to 70% of the chemical energy provided to the engine in the form of fuel is wasted, mostly from rejected heat. In a system that carries out work, heat loses are inevitable due to the first law of thermodynamics. The most arbitrary heat loses are that of the exhaust and cooling systems, but losses can also be on behalf of pumping losses, internal friction, drivetrain slippage, and other accessories. In fact, during in-town driving a typical vehicle will utilize, on average, only about 13% of the actual fuel energy to propel forward. To put it into context, a diagram presenting the energy pathway is located in Figure 2. Fortunately, developed technologies allow the conversion of a percentage of waste heat back to usable energy via several harvesting systems. Automobiles and especially heavy-duty commercial on and off highway vehicles have under scrutiny been lately [20,60,61].

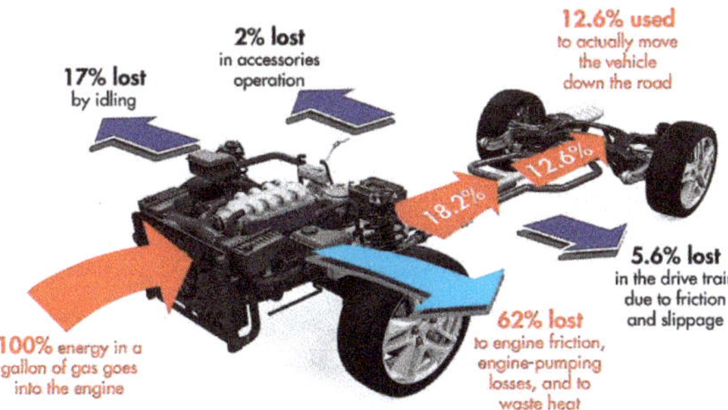

Figure 2. Typical city driving energy pathways.

In the US alone, tractor trailers, delivery vans, garbage trucks, and more are expected to reduce 25% of their exhaust emissions by 2027, with a potential to avoid up to 1.1 billion metric tons of carbon dioxide emissions. Many automotive industries that design and manufacture WHR systems seem to avoid mentioning the disadvantages of these promising devices. For example, they tend to incur an increase in exhaust backpressure, which has a direct impact on engine fuel consumption. Not many preliminary studies can be found that take exhaust backpressure fluctuation caused by the heat exchanger or turbine into account. As a result, the comprehension data, in consideration of designing and optimizing the components, is very limited. Another negative aspect is that integrated WHR units come alongside a weight penalty. The additional inertia will inevitably cause further fuel consumption and, subsequently, an increase in BSFC. These drawbacks are easy to neglect when heat harvesting percentages and thermal efficiencies of WHR means as well as the power unit are the center of attraction of the investigation [20,62,63].

1.2. Steam/Organic Rankine Cycle

The Rankine cycle (RC) is considered an ideal cycle for vapor power plants. It is comprised by four primary components, a pump, a boiler/evaporator, a power expansion turbine, and a condenser. The pump begins the cycle by pumping the working fluid through the system. The evaporator or boiler applies the recovered waste heat on the fluid, thus raising its temperature and pressure, creating (in some cases) superheated steam. The fluid is then expanded in the turbine utilizing the built up temperature and pressure by generating power through a shaft. The process is continued by the condenser, which condenses the vapor back to liquid form ready to be pumped again for another cycle [6,64,65].

The actual and ideal thermodynamic evaluation of the Rankine cycle operations can be slightly different from each other. In an ideal RC system, the compression and expansion processes in the pump and turbine, respectively, are considered isentropic. In an identical trend, the heat addition and heat rejection processes in the evaporator and condenser, respectively, are regarded as constant. This transliterates these processes as internally reversible. A reversible process is the process that can be reversed without leaving any traces to the environment or surroundings. Since the pump, evaporator, turbine, and condenser are all steady-flow components, the RC can be analyzed as a closed loop, steady-flow process following the steady-flow energy equation. This in turn implies that no heat engine will have a thermal efficiency of 100% because there must be a low temperature sink for the heat to be transferred to. In addition, the actual RC system suffers from losses throughout the systems cyclic function such as friction losses, piping losses, and heat transfer to surroundings. All of these losses cause an irreversible increase in entropy [2,9,10,32].

1.3. Organic Rankine Cycle and Internal Combustion Engines

The only difference between the conventional RC and the organic Rankine cycle (ORC) is the substance which circulates within the system. A traditional RC is known for using H_2O (water) as a working fluid. In the case of ORC, the operating fluid is an organic liquid element accompanied by a greater molecular mass and reduced boiling point compared to that of H_2O. These alternative working fluids can demonstrate a number of thermodynamic benefits as they allow the system to operate by using downsized temperatures. Consequently, the operation of the ORC results in greatly reduced thermal efficiency readings owning to the lower temperature transactions. However, this also has a positive impact on the total operational cost, as far less heat energy is required to produce a given power. The promising potential of low and moderate heat operation is the primary reason the ORC has grown to be popular amongst automobile industry research and development departments.

This WHR method, however, is not limited to ICE applications, but is also utilized by a number of other heat rejecting machinery, such as geothermal plans, solar thermal systems, and biomass plants, as well as industrial processes. The unstable transient and remarkably variable operational profiles of automotive vehicles make it more demanding to implement ORC systems and, therefore, the technology is expected to hit the market around 2020. This is mainly due to the non-existent control methods and instruments, which are of great necessity for the safety, performance, reliability, and durability of the power unit and ORC. Correspondingly, ORC patents are primarily developed and promoted for immobilized power production units as well as marine purposes [41,44,48,62]. In an investigation of an ORC system for a heavy-duty truck application and a passenger-car application, a 3.4% estimated reduction in fuel consumption was obtained. This was the result of a turbine efficiency of 58%, after the custom blade was designed on CFD software specifically for the employed ORC system. During the analysis of a diesel-Rankine cycle combination in a different HDDE case [58], it was concluded that during full load conditions (BMEP = 2 Mpa) a BSFC reduction of 2.6–3% was possible. The values of the temperature, pressure and, working fluid flow rate were all estimated by the thermodynamic characteristics of the multifarious utilized substances, namely, water (H_2O), methanol (MeOH), toluene (PhMe), pentafluoropropane (R245fa), and tetrafluoroethane (R134a). Moreover, the performance of an ORC system by integrating the WHR system on a 15 L diesel engine was investigated. The simulation assessment was performed by gathering the turbine shaft power from the exhaust enthalpy during steady-state operation. The use of dual finned heat exchangers with identical dimensions and properties was implemented in a parallel sequence after the turbocharger turbine to collect waste energy. The total power output of the engine was increased by an estimated 5%, while the engines' pumping losses were kept at a maximum total of 4 kW. It was also supported that much research based on WHR seem to neglect circumstantial disadvantages. Investigations based on ORC are no exception, with a couple of crucial unmentioned performance characteristics. These include the effect of refrigerant flow rate on ORC performance as well as the effects of pressure drops through the heat exchangers, with the resulting parasitic flow-work losses, two negative aspects which are overcompensated considering that the average theoretical integrated vehicle ORC system yields an increase in thermal efficiency of about 6–15%. As far as ORC fitment is concerned, depending on packaging and weight limitations of the given vehicle, the ORC recovery method can include multiple supplementary components. Typical layout implementations of ORC systems to HDDE can include twin parallel or in-series evaporators, individual for the exhaust and EGR valve. Furthermore, the introduction of a recuperator has found its way into the system due to the possibility of ORC efficiency increments. It is typically positioned between the turbine and the condenser and its functionality is to recuperate some of the heat before it is released to the heat sink by the condenser. Preheating the working substance with the aid of a charge air cooler (CAC) also constitutes an investigation possibility. Other researchers suggest replacing current engine block cooling techniques with ORC working fluids to take advantage of the additional waste heat and improve power regeneration. On the other hand, all of these efforts and aspects tend to increase the systems complexity rather than provide considerable ORC gains. Hence, a straightforward simplistic ORC composition is a more appealing solution for vehicle integration than most [6,40,44,65,66].

1.4. Engine Turbocompounding

In a conventional turbocharger, the engines' exhaust gas heat and airflow energy is harvested by a turbine, which is connected to a compressor through a common shaft. In the compressor side, air is induced and pressurized in the intake manifold, which increases the total power output of the engine with a small penalty on exhaust backpressure. A Turbo-compounding (T/C) system operates in a similar fashion, with the only difference being that there is no compressor at the end of the turbine shaft. The T/C turbine would be typically placed at the outlet of the primary turbine, therefore being driven by the leftover energy, translating it into a torque. T/C is a potentially prosperous WHR method, which can either be of a mechanical or electrical nature. A T/C system is a relatively less complicated arrangement compared to the ORC and this could potentially result in a lower unit production cost and a lighter component all together. On the other hand, one of the primary disadvantages of T/C implementation is the increment of engine backpressure and pumping losses, even more so than the ORC systems' heat exchanger. Exhaust backpressure is directly proportional to cylinder pumping loses. Hence, during meager engine speeds and loads, the total engine brake power output is prone to suffer.

As mentioned before, the power produced from the turbine can be manipulated either electrically or mechanically [10,14,30,67].

1.4.1. Electrical Turbo-Compounding

In an electric T/C system, an alternator/stator converts the turbines' rotational shaft power into electrical power. This electricity then either returns to the main battery to be stored for later usage or is immediately effective to operate various engine/vehicle components such as the starter motor, headlights, etc. Another possibility is the integration of an electrical compressor acting as a supercharger to assist the vehicles acceleration during low engine speeds where turbo-lag is yet to be overcome. Furthermore, the function of an electric motor directly mounted to the engines' crankshaft can also assist engine operation. Apart from throttle response, the techniques described can additionally improve fuel economy. In fact, in an electric T/C system, the estimated indication of reduction in fuel consumption ended up being a maximum of 10%. In addition, the strategy of the motor to crankshaft scheme seemed to enhance drivability and engine flexibility during transient periods. This in turn decreased exhaust gas emissions for an altogether greener engine activity. One of the other main advantages of electrical turbo-compounding is the space saving characteristics. Whether it is implemented as an integrated unit or a separate turbo-generator, it is a neat and tightly packaged component compared to the mechanical T/C system. On the other hand, the downside in the fitment of the electric T/C system is that it would generally require modifications to the existing turbomachinery [2,6,13,49,64]. This means that the already implemented turbocharger would have to be customized to incorporate the addition of a stator and rotor doublet in-between the turbine and the compressor impellers. However, there is the possibility that the electrical system is mounted on a separate turbine downstream of the main power turbine, also called a turbo-generator. This would diminish the need for existing turbo modifications because it is an independent and standalone unit.

1.4.2. Mechanical Turbo-Compounding

Similarly, the mechanical T/C system operates again by the addition of a secondary power turbine mounted sequentially after the principal turbine, to scavenge the surplus energy. The generated power is afterwards transmitted, via a shaft, through a gearbox unit followed by a mechanical coupling to the power units' crankshaft. It is generally a low volume and production cost system thus it is fundamentally applicable for medium and heavy-duty diesel power units. These leading HDDE manufacturers have been investigating the effects of different T/C arrangements with satisfactory results, namely a typical reduction in BSFC on a scale ranging between 3–6%. Investigations proved that the total improvement in incremental fuel consumption strictly due to the turbo-compounding action was a 4.2% to 5.3% estimate depended upon the terrain or mission load factor. Incorporating a mechanical T/C system in favor of an 11 Liter 6 cylinder turbocharged diesel engine resulted in a total of 5% reduction in BSFC during full load operation [42,43,45].

2. Engine Waste Heat Recovery System Modelling

In this section, the methods and tools used to assess the effects of a T/C system against the effects of an integrated ORC are explained. Important performance parameters, such as BSFC, thermal efficiency, power output, and overall fuel consumption are monitored and examined in conjunction with two WHR methods. The same virtual turbocharged HDDE was utilized for both the adopted techniques in favor of result accuracy. The T/C system was regarded as a secondary turbine arrangement downstream of the primary power turbine. With the aid of 1-D computer software (GT-Power), the engine, ORC, and T/C systems are modelled and optimized via trial and error simulations.

2.1. Engine Modelling and Calibration

An in-line, 6 cylinder, 11.5 liter, and turbocharged HDDE is assumed as a base research engine. The engine's major specifications are presented in Table 1. Figure 3 illustrates the engine model developed in GT-Power. The engine model is the same validated engine model as featured in Karvountzis-Kontakiotis et al [58], presented at the SAE World Congress. The model was further modified, in addition to ORC, to be able to simulate turbo-compounding operation.

Table 1. Modelled engine specifications.

Specification	Value
Engine Type	In-line 6, 4-Stroke, Diesel CI, Common Rail
Bore × Stroke	130 × 144 (mm)
Displacement	11.5 (L)
Compression Ratio	19:1
Max Power	478.3 [kW] @ 2500 RPM
Max Torque	1850 [N·m] @ 2050 RPM
BSFC at Peak Efficiency	214.7 [g/kW·h]
RPM Range	850–2600

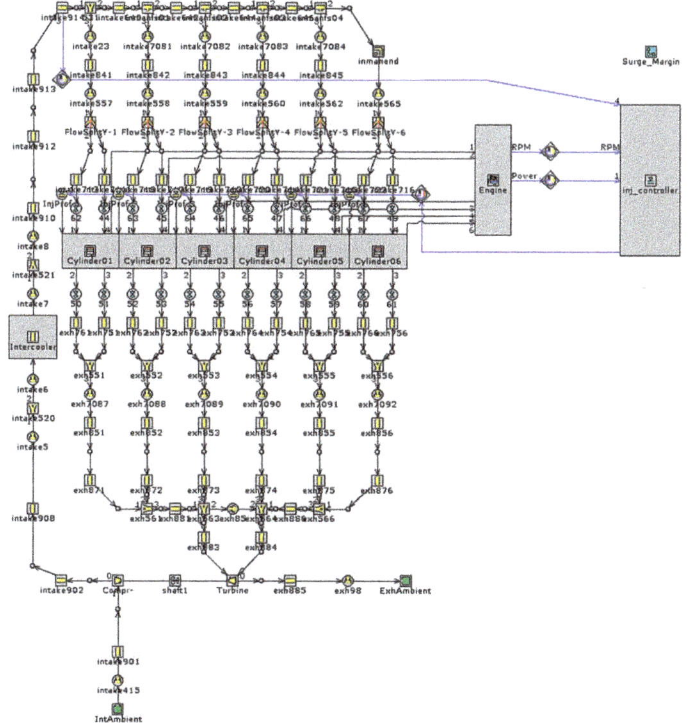

Figure 3. Engine model in GT-Power software.

A typical modern direct injection diesel engine is capable of working over a speed range of 600 rpm to 2600 rpm. This speed range is larger than normal but allows the data presented to be used over any part of that range. Therefore, the engine model was operated through 36 different cases ranging from 850 rpm to 2600 rpm using 50 rpm increments for a better result accuracy. It was decided that the best way to approach the optimization process was to limit the number of variables. This task proved to be challenging because, apart from engine speed, the operational cases varied in terms of engine load, fuel mass flow rate, power target of injector controller, turbine speed, etc.

A solid baseline of results was achieved by running the two WHR rivals at operating points, which the engine spends most of its working time. Figure 4 presents the typical engine speed and torque time percentage distribution for a crawler loader. The engine tested during this task is the type of engine that could be used for similar off-highway vehicles, as in Figure 4.

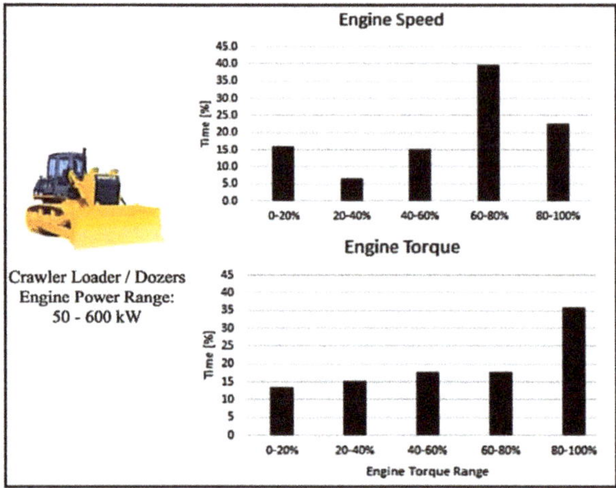

Figure 4. Typical HDDE time operational profiles.

In order to identify the ideal benchmarks in which the WHR systems will be constructed, optimized, and simulated, it is critical to run the engine simulation under real-life operating circumstances. In general, these optimum operational points are where the power unit produces certain usable benefits, such as maximum exhaust gas flow rates, lowest BSFC etc. Therefore, emphasis has been given to these points, which resemble an operation this engine is most likely to have under a typical working condition, similar to the example illustrated in Figure 4. Therefore, the WHR systems are assessed and compared for 6 dominant engine functioning points, which are determined by the calibration process. These include running at idling, ideal BSFC, part load, maximum torque output, maximum brake power output, and maximum exhaust flow rate. These are denoted by X1, X2, X3, X4, X5, and X6 respectively. As a gauge, the obtained baseline engine specifications are plotted using GT-Power 2-D graphical representations. This includes the BSFC contour map, which also indicates individual BMEP readings as well as power versus torque curves, all plotted in Figures 5 and 6 respectively. For the engine assessment, performance parameters, and comparison accuracy, both WHR methods are implemented on the same engine model.

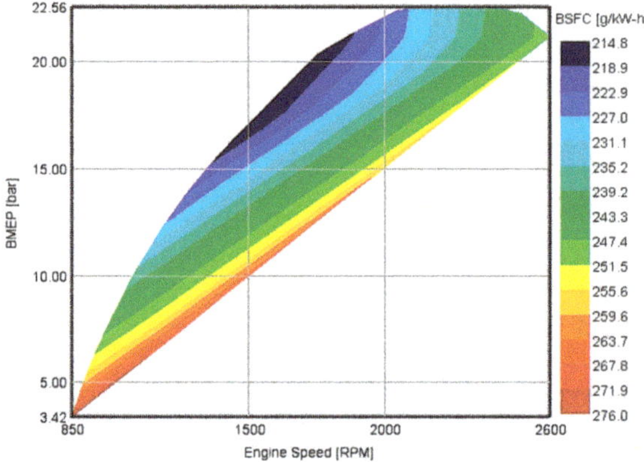

Figure 5. Baseline BSFC contour map.

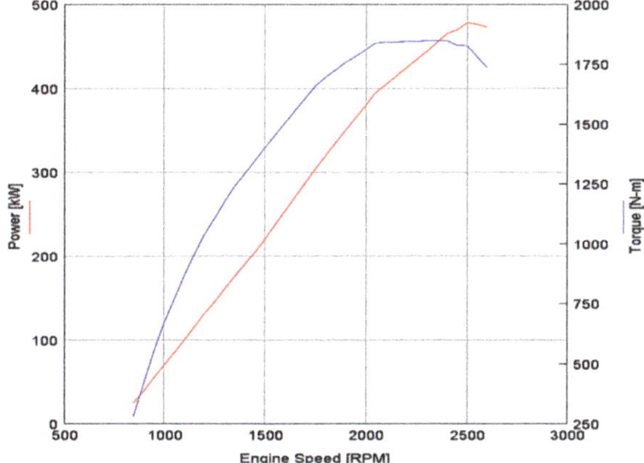

Figure 6. Baseline Power vs. Torque.

2.2. ORC System Modelling

The ORC system model is created and optimized in GT-Power software according to the predetermined aims and objectives of the study. The ORC model was kept to a minimal level and thus consisted of the following four main components: The pump, the boiler/evaporator, the expansion turbine, and the condenser. The pump and turbine elements were each coupled to a speed governor that sets the speed for each case run. These speed governors allowed turbine and pump speed variations for each operational point and, as explained later, this proved to be of significant value for the determination of the individual points' maximum performance enhancement. However, an increase in pump speed provokes an increase in work input requirement. Thus, to accomplish positive results, the energy recovered by the system will have to unavoidably reimburse the energy cost necessary for the pump operation.

The amount of energy deducted by the pump has a direct impact on the ORC system's efficiency due to the following relation. Another important factor which greatly interferes with the efficiency of the ORC system is the working fluid, so the refrigerant of type R245fa (Pentafluoropropane) is selected due to its advantageous low temperature heat recovery characteristics. The model of ORC system setup is illustrated in Figure 7. The control volume was considered to be adiabatic and, therefore, during all processes there was no heat escaping through the walls and surrounding features. This signifies that the exhaust gas pressure and temperature at the inlet and outlet of the evaporator were equal. Similarly, the coolant pressure and temperature at the inlet and outlet of the condenser are kept at an equal value. The next step is implementing the engines' turbocharger turbine outputs as ORC inputs at the heat exchanger. That includes the exhaust gas mass flow rate and temperature in the exhaust pipe downstream of the turbine for the six major running points of X1, X2, X3, X4, X5, and X6. Since the important aspect was to understand the strengths and weaknesses of each WHR system, the ORC pump and turbine speeds are altered during testing. Thus, the pump and turbine speeds were set according to literature review and benchmarking values were set, starting from 1500 rpm reaching up to 2500 rpm using 250 rpm increments. The design parameters of ORC used in the simulation are presented in Table 2.

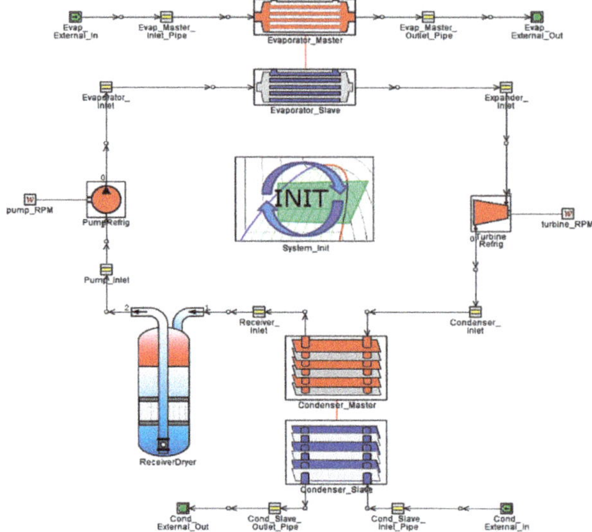

Figure 7. ORC system model in GT-Power.

Table 2. Component design parameters of ORC using R245fa refrigerant.

Design Parameters	ORC's Main Components					
	Evaporator (Exhaust)	Evaporator (Organic Fluid)	Condenser (Coolant)	Condenser (Organic Fluid)	Turbine Expander	Pump
Average Inlet Pressure (bar)	1.00102	24.9	2.15	3.28	24.3	2.6
Average Outlet Pressure (bar)	1	24.3	2	2.6	3.28	24.9
Average Pressure Drop (bar)	0.0010197	0.631	0.148264	0.674932	-	-
Average Inlet Temperature (K)	973.1	315.8	296.1	405.1	445.2	314.1
Average Outlet Temperature (K)	450.7	445.2	302.6	314.1	405.263	315.8
Average Mass Flow Rate (g/s)	140	269.2	3394.6	269.3	0.269	0.269
Combined Energy Rate out of Fluid (kW)	78.7	−78.7	−73.2	73.2	-	-
Average Map Pressure Ratio	-	-	-	-	7.37	-
Average Efficiency (%)	-	-	-	-	51.61	61.42
Average Power (kW)	-	-	-	-	5.3	0.75
Average Pressure Rise (bar)	-	-	-	-	-	22.3

2.3. Turbocompound System Modelling

Similar to the engine and ORC system model, the simulation of the T/C system model is performed with the aid of GT-Power software. On an industrial technicality level, the T/C system would have to be modelled on a separate template with the full extent of its integrated components. However, for this investigation and simplicity purposes, a simple secondary turbine is placed posterior to the primary turbine. In the pursuance of supervision and manipulation reasons, the turbine is incorporated with a rotational speed regulator as well as a signal output monitor. This model is a baseline calculation estimate and not a detailed representation of a T/C system. For example, the current T/C system model arrangement does not consider mechanical or electrical losses and thus the simulation will not represent real life expectation conclusively. The level of uncertainty is particularly higher at the developmental variables of the turbine, such as the performance map, which is a necessity for the validation of the modelled T/C system. As a result, literature review provides the essential assumptions and input specifications in order to avoid any potential errors. The model of the integrated T/C system is displayed in Figure 8.

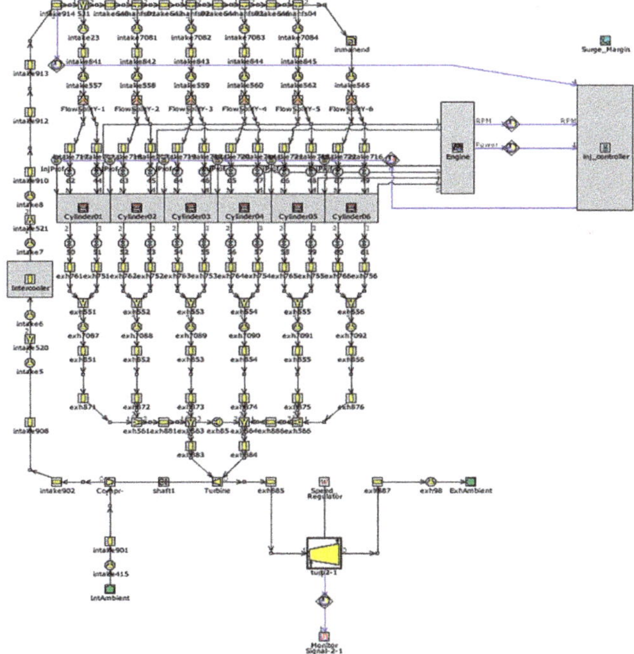

Figure 8. Turbo-compound system model.

Identically to the ORC system, benchmarking and literature review were not enough to optimize the T/C system model and thus it has to trail a series of experimental procedures by incorporating variable parameters as a plot of trial and error. In general, the power generated by the turbine will not be linear nor at its peak for all operating conditions. Therefore, the models' turbine is assessed during diverse rotational speeds, ranging from 20,000 rpm to 120,000 rpm using 10,000 rpm intervals, for all six benchmark points, resembling the process followed by the ORC system model. This will allow the identification of the optimum turbine speed for each given case and, hence, achieve maximum power output for the system as a total. The exhaust backpressure is expected to rise owing to the layout of the T/C system, which is placed directly after the turbocharger. A rise in exhaust gas pressure translates to further engine pumping losses during the intake and exhaust strokes. A comparable ORC system will also increase backpressure in the evaporator and this effect has been studied closely and published in a dedicated paper [58] by the Brunel University team. The important assumption here was that the heat exchanger technology, which can be employed, would have a minimal impact on fuel consumption. This is not an unfair assumption in view of the availability of heat exchanger technologies available with minimum impact to the gas exchange process. For the T/C case however, backpressure is heavily dependent on the operative expansion ratio and efficiency of the turbine expander—two parameters which are captured in the present investigation.

3. Results and Discussion

As mentioned, an in-depth investigation is conducted to assess and compare the advantages and disadvantages of integrated T/C and ORC systems. This section will provide a thorough comprehension for the results comparison of both WHR methods. In the specific engine property, between two configurations, the BSFC value ranges from a maximum value of 298.09 g/kW·h down to a minimum of 205.87 g/kW·h. The variation in power gains caused by the different pump and turbine speeds for the ORC and T/C systems (only the latter is varied for the T/C as there is no pump in the model) is also reviewed and explained in detail. The superior WHR method will be exclaimed by the BSFC reduction percentages and flexibility in operation.

3.1. Engine Waste Heat

The primary engine parameter, which is responsible for the performance of the WHR systems, is the available energy after the turbocharger turbine. The harvesting of waste heat is mainly depended on the accessibility of waste energy. The exhaust mass flow rate and exhaust gas temperature of the engine define the available energy for harvesting. By recording the exhaust mass flow rate and exhaust gas temperature values, one can determine the exhaust energy at the desired points of study. In general, more efficient energy gatherings are possible during top end operation. If the power unit is working at high engine speeds, the air mass inducted by the pistons increases, followed by larger amounts of fuel injected in the cylinders. Hence, the exhaust energy is enhanced due to the rise in exhaust gas mass flow rate. Therefore, it can be stated that there is a direct correlation between available exhaust energy, exhaust gas mass flow rate, engine speed, and possibly engine power output. Figure 9 represents the mass flow rate, BMEP, and engine speed, proportionality.

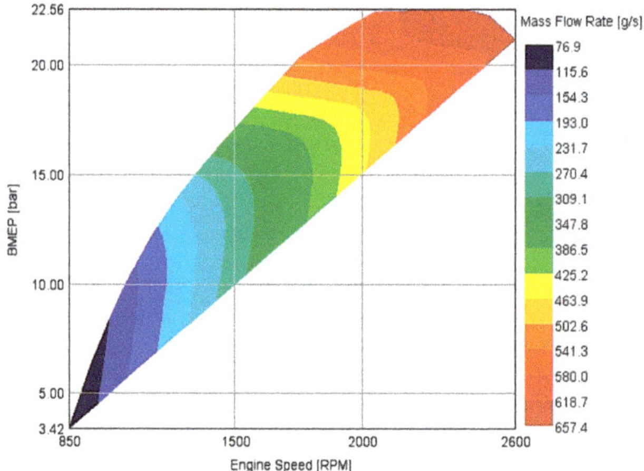

Figure 9. Variation of exhaust gas mass flow rate in accordance with engine speed and BMEP.

On the flipside, this correlation is not true for the exhaust gas temperature. In general, a high exhaust gas temperature signifies a deficient engine thermal efficiency. This is because a larger portion of energy is escaping from the combustion chamber, instead of being converted into usable mechanical work. It is observable from the contour plotting on Figure 10 that the exhaust gas temperature displays a higher temperature degree during intermediate BMEP and engine speeds.

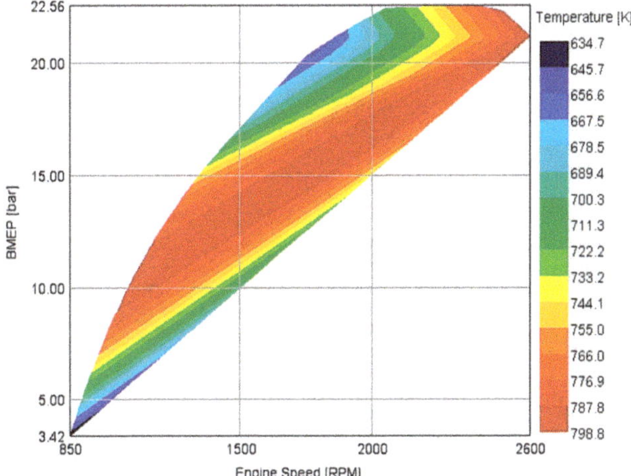

Figure 10. Variation of exhaust gas temperature in accordance with engines' speed and BMEP.

The lowest operating temperature level is achieved at the 1500 rpm point and around 20 bar of BMEP. This reveals the minimum BSFC value point (X2), which means that at that specific point, the engine is operating around peak thermal efficiency. Therefore, this proclaims that the exhaust temperature will predominately be higher at reduced thermal efficiencies. However, the contrast in exhaust gas temperature between the maximum and minimum thermal efficiency points is not of significant scale (approximately 160 °C). As a result, this sets the exhaust residue mass flow rate as the primary responsibility factor of regenerated power volume. This can also be validated by comparing the exhaust gas temperature contour map against that of the thermal efficiency profile, Figures 10 and 11, respectively.

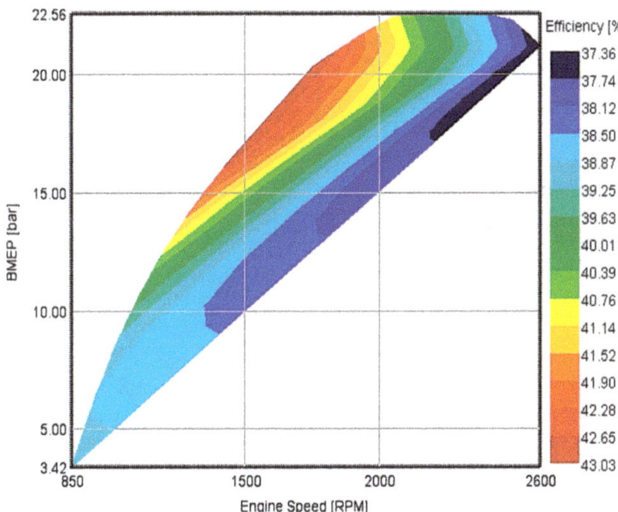

Figure 11. Variation of engine thermal efficiency in accordance with engines' speed and BMEP.

Conclusively, it can be suggested that there is a direct relation between WHR performance, engine speed, and reduced thermal efficiency profiles. This means that there is a greater potential to recover the engine's exhaust waste heat during operation at lower engine thermal efficiency.

3.2. Organic Rankine Cycle System

3.2.1. ORC System Speed Variation

By simulating the ORC system using various pump and turbine speeds it is possible to maximize the positive characteristics for each of the six assessment points. Figure 12 represents the system speed variation to the engine test points.

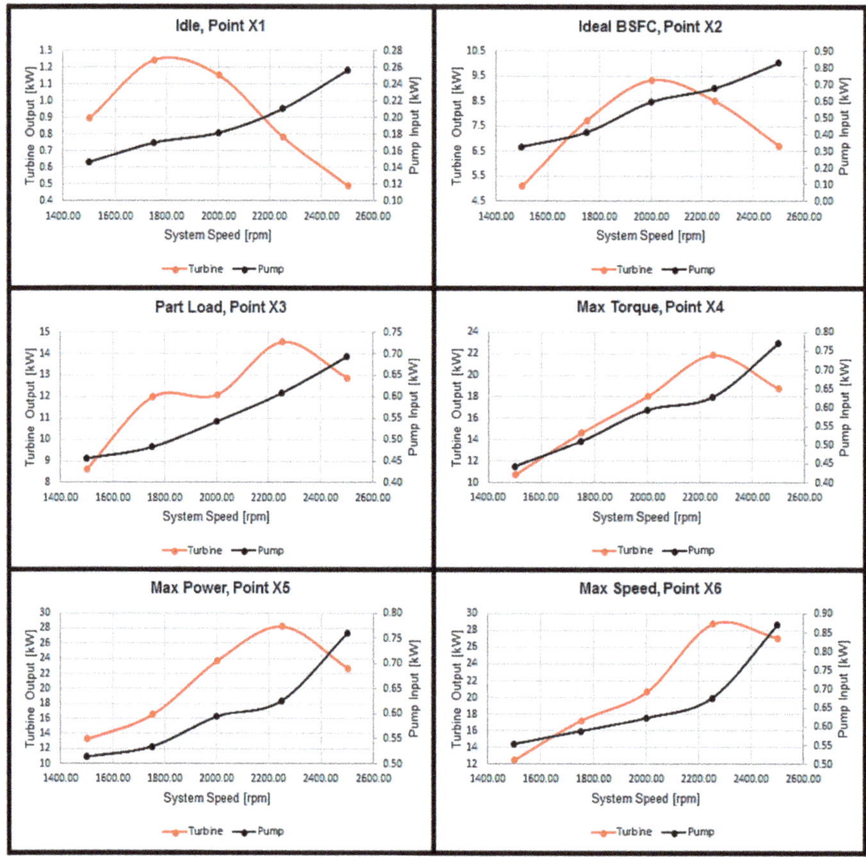

Figure 12. ORC system turbine power output in relation to system speed variation.

It is understandable that the power requirement to drive the pump increases proportionally with the pump speed. However, it is noticeable that the turbine power output fluctuates as the turbine and pump speed varies. On one hand, during the first two points (X1, X2), where the engine speed and thus exhaust gas mass flow rate is mediocre, the system is productive mostly at medium to low speeds. In addition, the variation in pump and turbine speed does not seem to provoke a considerable divergence in power output across the intervals. The reason is that the power output during low engine speeds is relatively low. On the other hand, as the engine speed rises (especially at points X5 and X6) so does the exhaust mass flow rate. Therefore, the amount of waste energy available for recovery increases. This achieves a greater power acquirement per working cycle and, hence, the difference between the power output levels between the system speed intervals grows significant. Table 3 includes the total range of pump and turbine speed performance variations for the ORC system, with the maximum work input and output values highlighted in red.

Table 3. ORC system speed variation simulations.

	X1		X2		X3	
Speed	Pump	Turbine	Pump	Turbine	Pump	Turbine
1500	0.146	0.894	0.328	5.112	0.456	8.631
1750	0.170	1.244	0.415	7.745	0.482	11.976
2000	0.182	1.152	0.596	9.339	0.543	12.104
2250	0.210	0.783	0.678	8.497	0.608	14.536
2500	0.256	0.490	0.828	6.723	0.693	12.863
	X4		X5		X6	
Speed	Pump	Turbine	Pump	Turbine	Pump	Turbine
1500	0.443	10.710	0.514	13.211	0.553	12.479
1750	0.510	14.641	0.534	16.483	0.588	17.217
2000	0.593	18.040	0.594	23.712	0.624	20.687
2250	0.628	21.843	0.626	28.225	0.676	28.749
2500	0.770	18.740	0.760	22.686	0.870	27.081

3.2.2. ORC System BSFC Reduction

For comparison purposes, all the calculations are conducted by utilizing the maximum values at each point. Now, BSFC is defined as the amount of fuel used per unit amount of power. By increasing power output, without increasing the fuel mass flow rate, the BSFC is reduced. Therefore, the addition of the engines' and ORC systems' power output would naturally decrease the BSFC value. Figure 13 shows the difference in BSFC.

Overall, the modelling and optimization of the ORC system managed to indicate a total average BSFC reduction of 4.78%, as explained in Table 4. The most substantial percentage value (5.6%) occurred at maximum engine operating speed point (X6). After the introduction of the ORC system, it can be observed that X2 is no longer the lowest BSFC point. A 4.36% BSFC reduction at X3 was enough to shift the ideal thermal efficiency benchmark.

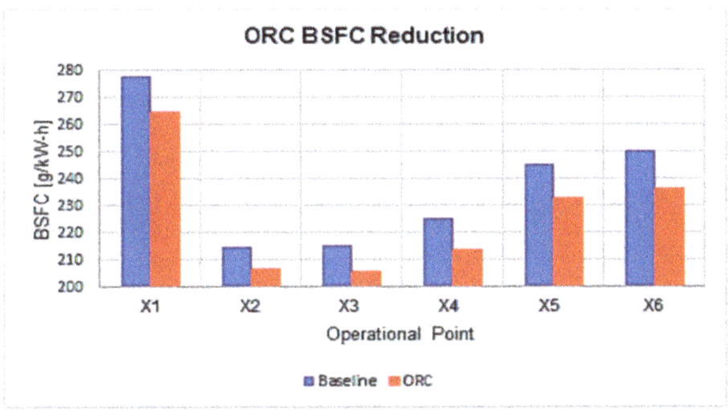

Figure 13. Reduction of BSFC in ORC system.

Table 4. Maximum BSFC reduction in ORC system.

Point	BSFC Reduction (%)
X1	4.64
X2	3.83
X3	4.36
X4	5.11
X5	5.16
X6	5.6
Total	4.78

The reduction in BSFC during system operation at point X2 presents a value of 3.83%, which is also the minimum reduction amount for the given engine. This means that the power unit at point X2 is already working at its peak thermal efficiency of approximately 43%, as earlier observed in Figure 11. Therefore, any further increments of this peak value are remarkably challenging to accomplish due to the reduced exhaust mass flow rate and temperature. Oppositely, the reason the BSFC reduction percentage is the greatest at X6 (5.6%) owns mostly to the inflated exhaust gas mass flow rate, which implements the largest impact as proved previously. Figure 14 illustrates the net power output to the BSFC reduction percentages.

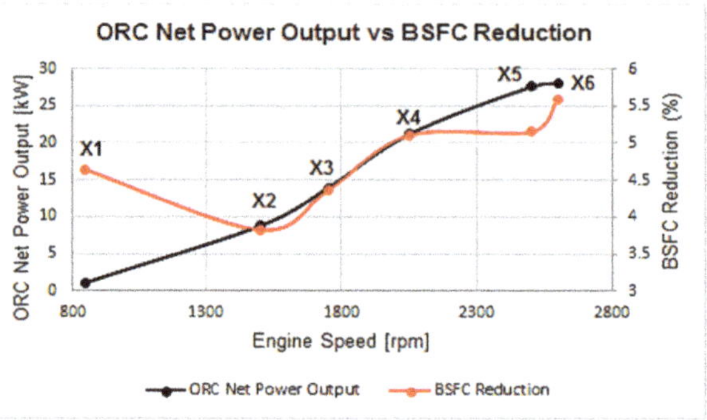

Figure 14. Variation of BSFC in accordance with ORC net power output.

The thermal efficiency of the ORC system model is plotted in Figure 15, wherein the thermal efficiency is calculated to be between 4% and 8%. It is not surprising that the thermal efficiency of the ORC system (η_{therm}) is at a very low mark considering that the pump and turbine were never designed to work in accordance with the specific engine exhaust outlets.

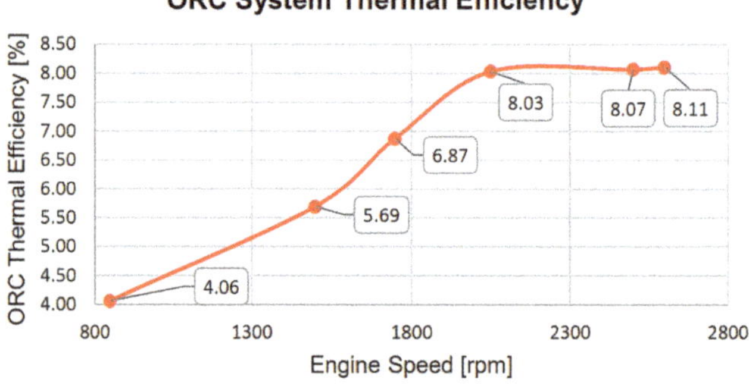

Figure 15. Thermal efficiency ORC system.

In fact, the thermal efficiency values would decrease even further. For example, if the system had not been configured as adiabatic, heat losses through the surroundings would, as a supplementary, encourage an even less efficient ORC operation. The low thermal efficiency of the ORC system contributed to the selection of the organic fluid, R245fa or any other for that matter. The decreased temperature input required for operation provokes the decrease in thermal efficiency. In the thermal efficiency graph, shown in Figure 15, there is an apparent inflation after the 2000 rpm mark. Nevertheless, despite the improved efficiency at high engine speeds, during high engine speeds and loads the ORC system is not able to take advantage of the additional and excessive amount of waste heat.

Engine efficiency can be seen to increase by a maximum of 5.69% at 1500 rpm. This is one of the speeds of interest for a heavy-duty engine, with the speed range of 1200–1500 rpm being, generally, the region of interest. The achieved efficiency compared well with engine data taken from the Brunel

University London engine ORC test facility and reported by Alshammari et al [68], at least from the point of view of cycle efficiency (4.3%) against a value of 4.95 to 5.69% in the region of interest in Figure 15.

3.3. Turbo-Compound System

3.3.1. Turbo-Compound System Speed Variation

Identical to the ORC systems' process, the T/C system is run through the six operating benchmarks (X1, X2, X3, X4, X5, and X6) which were defined during engine calibration. The power output of the T/C system model varies significantly with transitional turbine rotational speeds for a given operational occasion. It was revealed that the T/C system performed diversely for bottom, mid, and top range engine speeds. Figure 16 represents the ideal turbine speed configuration for each assessment point. During low fuel, mass flow rate, and load (point X1), the turbine was more productive by operating at high speeds of over 100,000 rpm. During testing at the highest thermal efficiency (point X2) and part load (point X3), a mediocre turbine speed was ideal, showing peak performance at 60,000 rpm. However, during top end runs (points X5 and X6) it is observable that high turbine speeds achieve the best power outputs. In particular, with a turbine speed setting of 100,000 rpm during max power output and a max engine speed (points X5, X6), the T/C system generates 34.66 kW and 35.68 kW respectively. The full extent of the variable speed turbine results is listed in Table 5. The peak turbine power outputs for each point are highlighted in red.

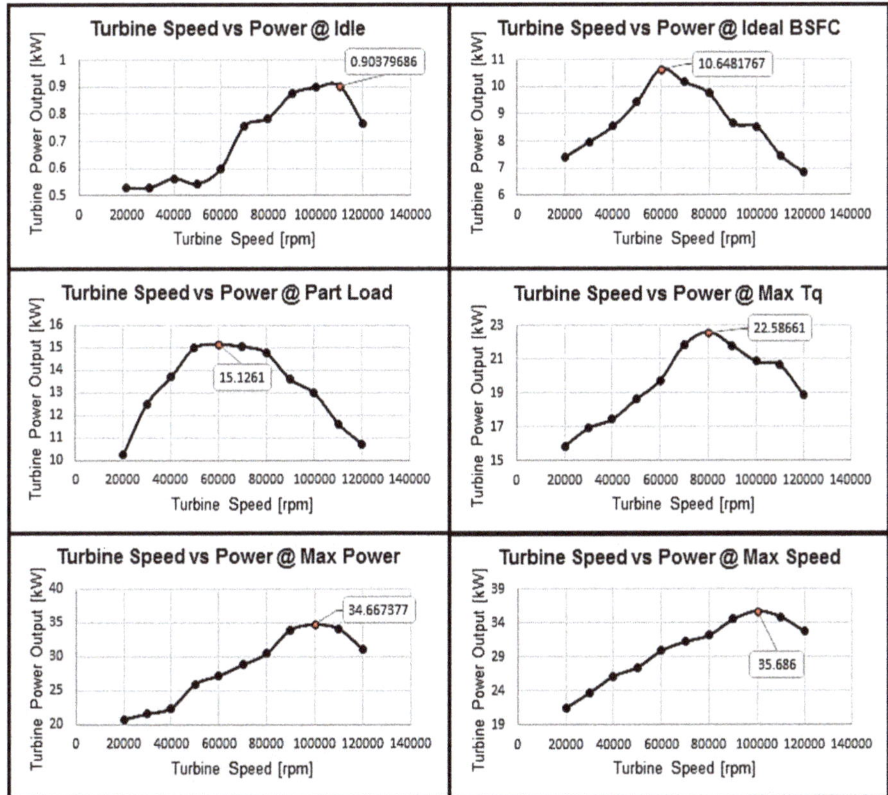

Figure 16. T/C system turbine power output in relation to turbine speed variation.

Table 5. Turbine analytical power output to speed variation.

Turbine Speed (rpm)	Idle (kW)	Ideal BSFC (kW)	Part Load (kW)	Max Torque (kW)	Max Power (kW)	Max Speed (kW)
20,000	0.530	7.409	10.265	15.874	20.762	21.520
30,000	0.530	7.964	12.480	16.933	21.635	23.719
40,000	0.564	8.537	13.683	17.452	22.451	26.148
50,000	0.544	9.426	14.973	18.638	25.976	27.452
60,000	0.598	10.648	15.126	19.747	27.160	29.897
70,000	0.756	10.156	15.033	21.805	28.799	31.211
80,000	0.784	9.754	14.770	22.587	30.425	32.146
90,000	0.874	8.647	13.589	21.770	33.916	34.524
100,000	0.897	8.504	12.981	20.836	34.667	35.686
110,000	0.904	7.468	11.627	20.620	34.037	34.827
120,000	0.765	6.832	10.725	18.869	31.042	32.684

3.3.2. Turbo-Compound System BSFC Reduction

Similar to the ORC system, the comparison purposes require the use of optimum power values, despite the fact that the yields are obtained using different turbine speeds. In addition, the BSFC difference between the baseline and turbo-compound engine is calculated. There is an important difference in the analysis of the T/C and ORC systems. The ORC system is modelled on a separate template whereas the T/C system is placed and assessed directly on the stock engine model as an integrated unit. As explained during the simulation section, due to the incorporation of the secondary turbine, the exhaust backpressure inflated, causing additional engine pumping losses. Figure 17 shows the escalation of exhaust backpressure after the introduction of the T/C system.

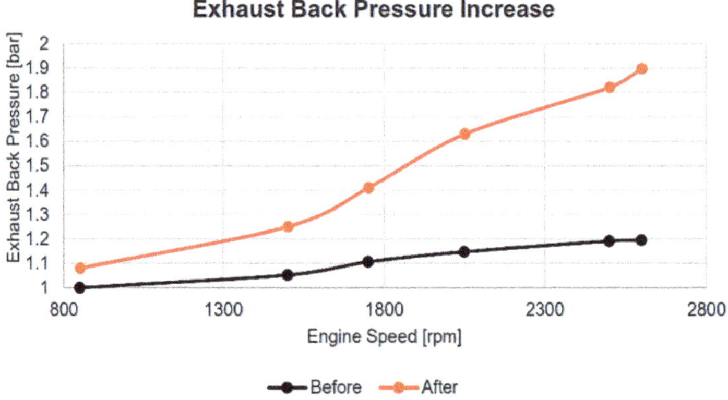

Figure 17. T/C system backpressure increment.

As expected by the increase in exhaust backpressure, the pumping losses are reflected by a proportional increase in BSFC. In fact, if the generated power output from the T/C system is not taken into account for the calculation of BSFC, the escalation in backpressure alone is enough to deteriorate the BSFC by a total average of almost 5.5% at low engine speeds. However, it is worth mentioning that this deterioration is, mostly, only distinguishable during low engine speeds, specifically as listed in Table 6. The results show an average of 3% increase in BFSC before applying the positive aspects of the T/C system in the standard engine model and, thus, BSFC is decreased, due to the harvested power. It is observable that this increase in BSFC is significant from point X1 to point X3 at an average of 5.5%. On the flipside, the BSFC values for points X4, X5, and X6 remain identical, with an average difference of less than 0.5%. This relation suggests that the higher exhaust backpressure only affects the engine thermal efficiency at low engine speeds. With an overall consideration, the effects of the T/C system on the engine are represented in Figure 18.

Table 6. T/C system analytical BSFC increments.

Engine Speed (rpm)	BSFC Increase (%)
850	6.864001
1500	5.270832
1750	4.345699
2050	0.554805
2500	0.468826
2600	0.536313
Total	3

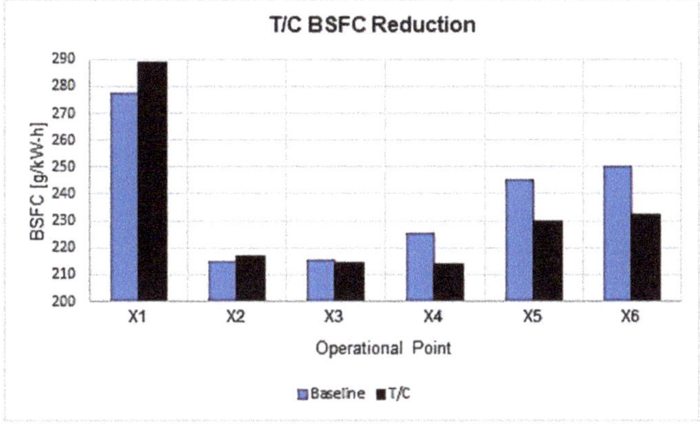

Figure 18. T/C system BSFC reduction.

It is notable that, in excess of 2000 rpm, the effect of exhaust backpressure on BSFC was essentially annihilated. After that moment, the T/C system had the opportunity to commence with the positive characteristics of its nature. The beneficial aspects are also evident when plotting the secondary turbines' power output with the BSFC reduction percentage, plotted on Figure 19. Notice that the BSFC reduction axis has a minimum value of zero. This is done to highlight the major benefits of the T/C system during top end operation, which touch a maximum value of 7%. Table 7 includes the specific reduction values at each operational point.

Identically to the ORC system, the point of maximum engine thermal efficiency, the lowest BSFC point, is shifted to the right-hand side, namely, from the minimum point, X2 (214.773 g/kW·h), to previous point of maximum torque output, X4 (213.884 g/kW·h). Major BSFC reductions are shown after point X4, meaning that if the given engine spends most of its operating time at high speeds the T/C system would have a theoretical reduction in BSFC at an approximate average of 6.5%.

Table 7. T/C analytical BSFC reduction.

Operational Point	BSFC Reduction (%)
X1	−4.02734
X2	−1.08217
X3	0.392548
X4	5.113118
X5	6.355747
X6	7.011852
Total	2.293959167

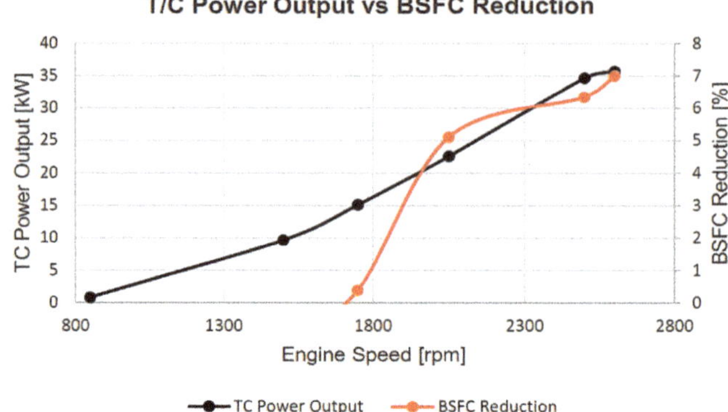

Figure 19. T/C power output vs. BSFC reduction.

3.3.3. Turbo-Compound System Efficiency

The thermal efficiency of the T/C system follows a similar fashion to the ORC system. Therefore, it is calculated by the amount of power gained over the exhaust energy input in the form of surplus heat. However, unlike the work input necessity for the ORC systems' pump, there was no work input required in the T/C system. As far as the exhaust energy is concerned, it was again calculated by the product of exhaust gas specific heat, mass flow rate, and temperature difference between the secondary turbines' inlet and outlet. Figure 20 illustrates the T/C systems' thermal efficiency plot.

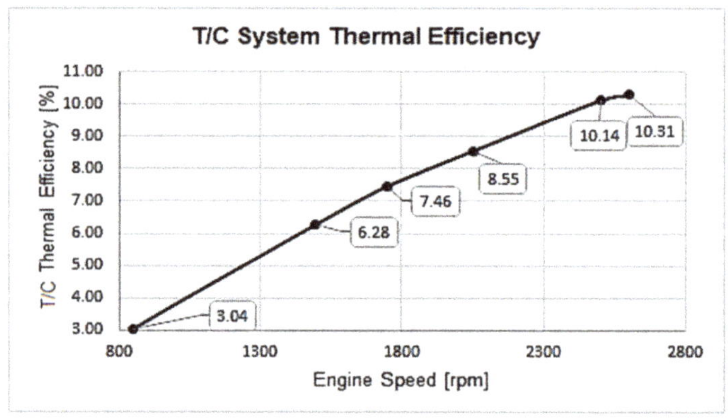

Figure 20. T/C system thermal efficiency.

A maximum of 10.31% of thermal efficiency may not seem significant, but it is heat that would otherwise be wasted. In addition, unlike the ORC system, of which the efficiency remains relatively stationary after the 2000-rpm mark, the T/C systems shows a continuously progressive increase.

3.4. ORC T/C System vs. ORC System Comparison

By assessing the two WHR methods individually, a solid base of strength and weakness points was set. It was noticeable that at point X6, which is the point for maximum engine speed, both WHR methods were at their peak BSFC reduction percentage. In fact, both produced maximum power output and reached maximum thermal efficiencies at point X6. This is due to the benchmarks' higher exhaust mass flow rate. On the other hand, the increment of exhaust mass flow rate subsequently reduced the exhaust gas temperature. As engine speed was increasing for a given engine load, fuel mass flow rate was also raised. At the same time, the compressor was forcing additional air into the cylinders.

As a result, the combustion process improved due to the higher oxygen content within the combustion chamber, featured by the increased air mass flow rate and air to fuel ratio. The improved combustion converts the chemical energy supplied by the diesel to effective mechanical work more effectively, instead of wasting it as energy in the form of heat. However, the reduction in BSFC was naturally greater at higher BSFC values because they are susceptible to permit more room for improvement. For the same reasons, it was observed that for both the T/C and ORC systems, the lowest BSFC point is conveyed after the WHR implementations.

Despite the excessive turbulent flow that undermines the already unstable exhaust gas pulse at the measuring point (after the turbine), the temperature of the exhaust gas demonstrated an adequate stability, which further benefits the WHR methods. For a clearer contrast resolution, the BSFC values from points X1 to X6 are plotted on the same graph for both the ORC and T/C systems while using the original plot as a gauge, as shown in Figure 21.

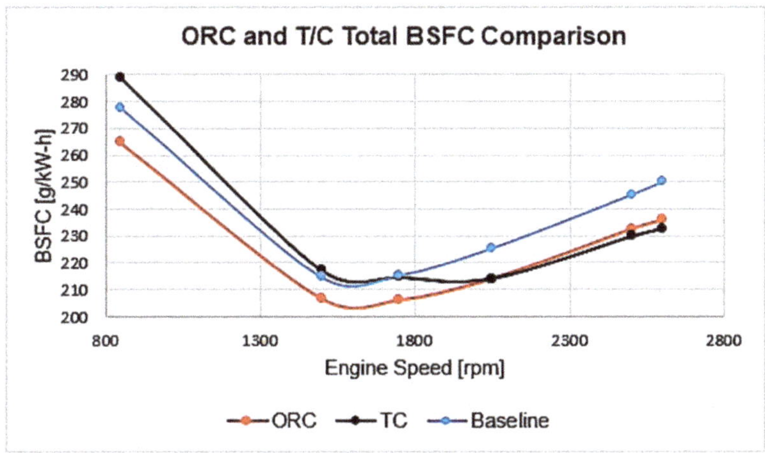

Figure 21. Comparison of total BSFC between ORC and T/C in relation to engines' speed.

The results indicated that the ORC system was more favorable during low engine speed operation. This is because the secondary turbine in the T/C system cannot produce enough mechanical power to compensate for the additional exhaust backpressure during low-end operation. However, it was a different story when the engine was running at high speed. At points X4, X5, and X6, the exhaust backpressure did not have a major impact on the BSFC value. In addition, the turbine thrived, due to the increased exhaust mass flow rate, and the power gain was in excess of 35 kW at the finishing point. That was not only enough to compensate for the low engine speed losses, but also to further decrease the overall BSFC by an average of 2.3%. The comparison of BSFC between the two WHR systems as well as the baseline engine is also visualized in Figure 22. As far as the physical mass and volume properties of the two WHR systems are concerned, both are at approximately the same level. A mechanical T/C system would naturally result in more weight, due to the incorporated gearbox unit.

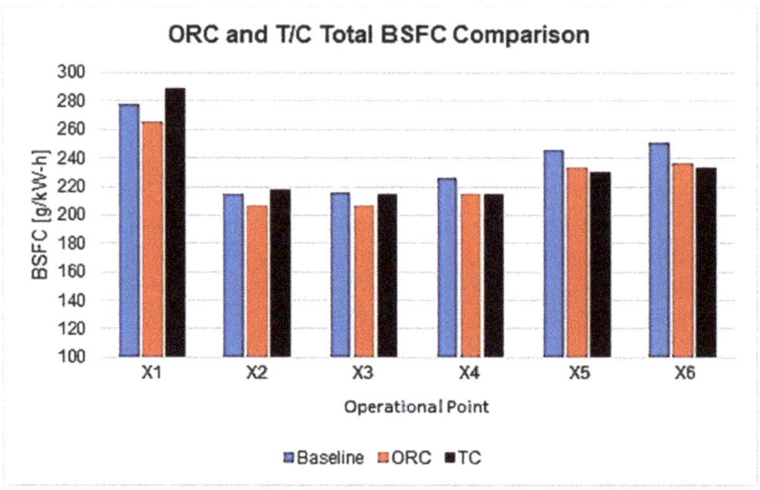

Figure 22. Comparison of total BSFC between ORC and T/C in relation to six individual points.

4. Conclusions

A comparison between two WHR systems using 1-D engine code simulation in GT-Power was conducted. The main aspect considered was their ability to reduce BSFC of an experimental engine model. To recapitulate, the effects of each systems' integration are listed below, as follows:

- The capability to regenerate power is determined by the availability of exhaust energy at the systems inlet conditions. That availability is reduced at ideal BSFC regions and increased during top end engine speeds.
- Exhaust energy is directly proportional to exhaust gas temperature and mass flow rate, however only the latter administers the greatest impact; temperature remains relatively unchanged across the engines' range.
- The power outputs for both WHR methods varied with the systems operational speed. The maximum power obtain by the ORC and T/C systems were 27.3 kW and 35.6 kW respectively.
- The ORC managed a total average BSFC reduction of 4.8%, whereas the T/C yielded an average of 2.3%.
- The thermal efficiencies of the ORC and T/C systems were considerably low, at max values of 8% and 10% respectively.
- The raise in exhaust backpressure by the T/C system affected low speed BSFC severely so much so that the system was unable to regenerate enough power to compensate for the additional fuel consumption.

The ORC system provided a more consistent WHR method, with progressive improvements in fuel consumption across the engines' speed range. However, the T/C system presents incomparable contributions in fuel economy during high-speed engine operation.

Author Contributions: A.M.A. and A.P. have written the paper context and performed the simulation works alongside the results presentation. V.E., A.S.-Z., and A.H. have carried out the design of experiment in the simulations.

Funding: This research received no external funding.

Conflicts of Interest: The authors declare no conflict of interest.

References

1. Karvountzis-Kontakiotis, A.; Mahmoudzadeh Andwari, A.; Pesyridis, A.; Russo, S.; Tuccillo, R.; Esfahanian, V. Application of Micro Gas Turbine in Range-Extended Electric Vehicles. *Energy* **2018**, *147*, 351–361. [CrossRef]
2. Shu, G.-Q.; Yu, G.; Tian, H.; Wei, H.; Liang, X. *Simulations of a Bottoming Organic Rankine Cycle (ORC) Driven by Waste Heat in a Diesel Engine (DE)*; SAE International: Warrendale, PA, USA, 2013.
3. Sprouse, C., III; Depcik, C. Organic Rankine Cycles with Dry Fluids for Small Engine Exhaust Waste Heat Recovery. *SAE Int. J. Altern. Powertrains* **2013**, *2*, 96–104. [CrossRef]
4. Teng, H.; Regner, G.; Cowland, C. *Waste Heat Recovery of Heavy-Duty Diesel Engines by Organic Rankine Cycle Part I: Hybrid Energy System of Diesel and Rankine Engines*; SAE International: Warrendale, PA, USA, 2007.
5. Yang, F.; Zhang, H.; Yu, Z.; Wang, E.; Meng, F.; Liu, H.; Wang, J. Parametric optimization and heat transfer analysis of a dual loop ORC (organic Rankine cycle) system for CNG engine waste heat recovery. *Energy* **2017**, *118*, 753–775. [CrossRef]
6. Zhang, X.; Mi, C. *Vehicle Power Management; Modeling, Control and Optimization*; Springer: London, UK, 2011; p. 346.
7. Mahmoudzadeh Andwari, A.; Said, M.F.M.; Aziz, A.A.; Esfahanian, V.; Salavati-Zadeh, A.; Idris, M.A.; Perang, M.R.M.; Jamil, H.M. Design, Modeling and Simulation of a High-Pressure Gasoline Direct Injection (GDI) Pump for Small Engine Applications. *J. Mech. Eng.* **2018**, *SI 6*, 107–120.
8. Said, M.F.M.; Aziz, A.B.A.; Latiff, Z.A.; Mahmoudzadeh Andwari, A.; Soid, S.N.M. *Investigation of Cylinder Deactivation (CDA) Strategies on Part Load Conditions*; SAE Technical Paper 2014-01-2549; SAE International: Warrendale, PA, USA, 2014.
9. Yamaguchi, T.; Aoyagi, Y.; Osada, H.; Shimada, K.; Uchida, N. BSFC Improvement by Diesel-Rankine Combined Cycle in the High EGR Rate and High Boosted Diesel Engine. *SAE Int. J. Engines* **2013**, *6*, 1275–1286. [CrossRef]
10. Yang, Y.; Zhang, H.; Xu, Y.; Zhao, R.; Hou, X.; Liu, Y. Experimental study and performance analysis of a hydraulic diaphragm metering pump used in organic Rankine cycle system. *Appl. Therm. Eng.* **2018**, *132*, 605–612. [CrossRef]
11. Zhang, J.; Zhang, H.; Yang, K.; Yang, F.; Wang, Z.; Zhao, G.; Liu, H.; Wang, E.; Yao, B. Performance analysis of regenerative organic Rankine cycle (RORC) using the pure working fluid and the zeotropic mixture over the whole operating range of a diesel engine. *Energy Convers. Manag.* **2014**, *84*, 282–294. [CrossRef]
12. Zhang, X.; Zeng, K.; Bai, S.; Zhang, Y.; He, M. *Exhaust Recovery of Vehicle Gasoline Engine Based on Organic Rankine Cycle*; SAE International: Warrendale, PA, USA, 2011.
13. Zhou, F.; Joshi, S.N.; Rhote-Vaney, R.; Dede, E.M. A review and future application of Rankine Cycle to passenger vehicles for waste heat recovery. *Renew. Sustain. Energy Rev.* **2017**, *75*, 1008–1021. [CrossRef]
14. Bell, C. *Maximum Boost: Designing, Testing and Installing Turbocharger Systems*; Robert Bentley, Incorporated: Cambridge MA, USA, 1997.
15. Bin Mamat, A.M.I.; Martinez-Botas, R.F.; Rajoo, S.; Hao, L.; Romagnoli, A. Design methodology of a low pressure turbine for waste heat recovery via electric turbocompounding. *Appl. Therm. Eng.* **2016**, *107*, 1166–1182. [CrossRef]
16. Boretti, A. *Improving the Efficiency of Turbocharged Spark Ignition Engines for Passenger Cars through Waste Heat Recovery*; SAE International: Warrendale, PA, USA, 2012.
17. Ghanaati, A.; Said, M.F.M.; Mat Darus, I.Z.; Mahmoudzadeh Andwari, A. A New Approach for Ignition Timing Correction in Spark Ignition Engines Based on Cylinder Tendency to Surface Ignition. *Appl. Mech. Mater.* **2016**, *819*, 272–276. [CrossRef]
18. Mahmoudzadeh Andwari, A.; Aziz, A.A.; Said, M.F.M.; Esfahanian, V.; Latiff, Z.A.; Said, S.N.M. Effect of internal and external EGR on cyclic variability and emissions of a spark ignition two-stroke cycle gasoline engine. *J. Mech. Eng. Sci.* **2017**, *11*, 3004–3014. [CrossRef]
19. Mahmoudzadeh Andwari, A.; Said, M.F.M.; Aziz, A.A.; Esfahanian, V.; Baker, M.R.A.; Perang, M.R.M.; Jamil, H.M. A Study on Gasoline Direct Injection (GDI) Pump System Performance using Model-Based Simulation. *J. Soc. Automot. Eng. Malays.* **2018**, *2*, 14–22.
20. Zhou, L.; Tan, G.; Guo, X.; Chen, M.; Ji, K.; Li, Z.; Yang, Z. *Study of Energy Recovery System Based on Organic Rankine Cycle for Hydraulic Retarder*; SAE International: Warrendale, PA, USA, 2016.

21. Arsie, I.; Cricchio, A.; Pianese, C.; Ricciardi, V.; De Cesare, M. *Modeling and Optimization of Organic Rankine Cycle for Waste Heat Recovery in Automotive Engines*; SAE International: Warrendale, PA, USA, 2016.
22. Bell, A.G. *Forced Induction Performance Tuning*; Haynes: Bristol, UK, 2002.
23. Cipollone, R.; Battista, D.D.; Gualtieri, A. Turbo compound systems to recover energy in ICE. *Int. J. Eng. Innov. Technol.* **2013**, *3*, 249–257.
24. Ghanaati, A.; Mat Darus, I.Z.; Farid, M.; Said, M.; Mahmoudzadeh Andwari, A. A Mean Value Model for Estimation of Laminar and Turbulent Flame Speed in Spark-Ignition Engine. *Int. J. Automot. Mech. Eng. Online* **2015**, *11*, 2229–8649. [CrossRef]
25. Mahmoudzadeh Andwari, A.; Azhar, A.A. Homogenous Charge Compression Ignition (HCCI) Technique: A Review for Application in Two-Stroke Gasoline Engines. *Appl. Mech. Mater.* **2012**, *165*, 53–57.
26. Chen, T.; Zhuge, W.; Zhang, Y.; Zhang, L. A novel cascade organic Rankine cycle (ORC) system for waste heat recovery of truck diesel engines. *Energy Convers. Manag.* **2017**, *138*, 210–223. [CrossRef]
27. Dolz, V.; Novella, R.; García, A.; Sánchez, J. HD Diesel engine equipped with a bottoming Rankine cycle as a waste heat recovery system. Part 1: Study and analysis of the waste heat energy. *Appl. Therm. Eng.* **2012**, *36*, 269–278. [CrossRef]
28. El Chammas, R.; Clodic, D. *Combined Cycle for Hybrid Vehicles*; SAE International: Warrendale, PA, USA, 2005.
29. Cochran, D.L. *Working Fluids for High Temperature, Rankine Cycle, Space Power Plants*; SAE International: Warrendale, PA, USA, 1961.
30. Mahmoudzadeh Andwari, A.; Pesiridis, A.; Esfahanian, V.; Salavati-Zadeh, A.; Karvountzis-Kontakiotis, A.; Muralidharan, V. A Comparative Study of the Effect of Turbocompounding and ORC Waste Heat Recovery Systems on the Performance of a Turbocharged Heavy-Duty Diesel Engine. *Energies* **2017**, *10*, 1087. [CrossRef]
31. Ringler, J.; Seifert, M.; Guyotot, V.; Hübner, W. Rankine Cycle for Waste Heat Recovery of IC Engines. *SAE Int. J. Engines* **2009**, *2*, 67–76. [CrossRef]
32. Serrano, J.R.; Dolz, V.; Novella, R.; García, A. HD Diesel engine equipped with a bottoming Rankine cycle as a waste heat recovery system. Part 2: Evaluation of alternative solutions. *Appl. Therm. Eng.* **2012**, *36*, 279–287. [CrossRef]
33. Lodwig, E. *Performance of a 35 HP Organic Rankine Cycle Exhaust Gas Powered System*; SAE International: Warrendale, PA, USA, 1970.
34. Mahmoudzadeh Andwari, A.; Pesyridis, A.; Esfahanian, V.; Said, M.F.M. Combustion and Emission Enhancement of a Spark Ignition Two-Stroke Cycle Engine Utilizing Internal and External Exhaust Gas Recirculation Approach at Low-Load Operation. *Energies* **2019**, *12*, 609. [CrossRef]
35. Matthew Read, I.S. Nikola Stosic and Ahmed Kovacevic, Comparison of Organic Rankine Cycle Systems under Varying Conditions Using Turbine and Twin-Screw Expanders. *Energies* **2016**, *9*, 614. [CrossRef]
36. Mavrou, P.; Papadopoulos, A.I.; Seferlis, P.; Linke, P.; Voutetakis, S. Selection of working fluid mixtures for flexible Organic Rankine Cycles under operating variability through a systematic nonlinear sensitivity analysis approach. *Appl. Therm. Eng.* **2015**, *89*, 1054–1067. [CrossRef]
37. Allouache, A.; Leggett, S.; Hall, M.J.; Tu, M.; Baker, C.; Fateh, H. Simulation of Organic Rankine Cycle Power Generation with Exhaust Heat Recovery from a 15 liter Diesel Engine. *SAE Int. J. Mater. Manf.* **2015**, *8*, 227–238. [CrossRef]
38. Çengel, Y.A. *Introduction to Thermodynamics and Heat Transfer*; McGraw-Hill: New York, NY, USA, 2007.
39. Mahmoudzadeh Andwari, A.; Aziz, A.A.; Muhamad Said, M.F.; Abdul Latiff, Z. Controlled Auto-Ignition Combustion in a Two-Stroke Cycle Engine Using Hot Burned Gases. *Appl. Mech. Mater.* **2013**, *388*, 201–205. [CrossRef]
40. Pesiridis, A. *Automotive Exhaust Emissions and Energy Recovery*; Nova Science Publishers, Incorporated: Hauppauge, NY, USA, 2014.
41. Mahmoudzadeh Andwari, A.; Pesiridis, A.; Karvountzis-Kontakiotis, A.; Esfahanian, V. Hybrid electric vehicle performance with organic rankine cycle waste heat recovery system. *Appl. Sci.* **2017**, *7*, 437. [CrossRef]
42. Shu, G.; Zhao, J.; Tian, H.; Wei, H.; Liang, X.; Yu, G.; Liu, L. *Theoretical Analysis of Engine Waste Heat Recovery by the Combined Thermo-Generator and Organic Rankine Cycle System*; SAE International: Warrendale, PA, USA, 2012.
43. Tennant, D.W.H.; Walsham, B.E. *The Turbocompound Diesel Engine*; SAE International: Warrendale, PA, USA, 1989.

44. Wang, E.; Yu, Z.; Zhang, H.; Yang, F. A regenerative supercritical-subcritical dual-loop organic Rankine cycle system for energy recovery from the waste heat of internal combustion engines. *Appl. Energy* **2017**, *190*, 574–590. [CrossRef]

45. Wilson, D.E. *The Design of a Low Specific Fuel Consumption Turbocompound Engine*; SAE International: Warrendale, PA, USA, 1986.

46. Mahmoudzadeh Andwari, A.; Aziz, A.A.; Said, M.F.M.; Latiff, Z.A. A Converted Two-Stroke Cycle Engine for Compression Ignition Combustion. *Appl. Mech. Mater.* **2014**, *663*, 331–335. [CrossRef]

47. Kolasiński, P.; Błasiak, P.; Rak, J. Experimental and Numerical Analyses on the Rotary Vane Expander Operating Conditions in a Micro Organic Rankine Cycle System. *Energies* **2016**, *9*, 606. [CrossRef]

48. Quoilin, S.; Broek, M.V.D.; Declaye, S.; Dewallef, P.; Lemort, V. Techno-economic survey of Organic Rankine Cycle (ORC) systems. *Renew. Sustain. Energy Rev.* **2013**, *22*, 168–186. [CrossRef]

49. Katsanos, C.O.; Hountalas, D.T.; Pariotis, E.G. Thermodynamic analysis of a Rankine cycle applied on a diesel truck engine using steam and organic medium. *Energy Convers. Manag.* **2012**, *60*, 68–76. [CrossRef]

50. Kölsch, B.; Radulovic, J. Utilisation of diesel engine waste heat by Organic Rankine Cycle. *Appl. Therm. Eng.* **2015**, *78*, 437–448. [CrossRef]

51. Kulkarni, K.; Sood, A. *Performance Analysis of Organic Rankine Cycle (ORC) for Recovering Waste Heat from a Heavy Duty Diesel Engine*; SAE International: Warrendale, PA, USA, 2015.

52. Kunte, H.; Seume, J. *Partial Admission Impulse Turbine for Automotive ORC Application*; SAE International: Warrendale, PA, USA, 2013.

53. Lion, S.; Michos, C.N.; Vlaskos, I.; Rouaud, C.; Taccani, R. A review of waste heat recovery and Organic Rankine Cycles (ORC) in on-off highway vehicle Heavy Duty Diesel Engine applications. *Renew. Sustain. Energy Rev.* **2017**, *79*, 691–708. [CrossRef]

54. Mahmoudzadeh Andwari, A.; Aziz, A.A.; Said, M.F.M.; Latiff, Z.A.; Ghanaati, A. Influence of Hot Burned Gas Utilization on the Exhaust Emission Characteristics of a Controlled Auto-Ignition Two-Stroke Cycle Engine. *Int. J. Automot. Mech. Eng. Online* **2015**, *11*, 2229–8649.

55. Hopmann, U.; Algrain, M.C. *Diesel Engine Electric Turbo Compound Technology*; SAE International: Warrendale, PA, USA, 2003.

56. Heywood, J.B. *Internal Combustion Engine Fundamentals*; McGraw-Hill: New York, NY, USA, 1988.

57. Nicolas Stanzel, T.S. Markus Preißinger and Dieter Brüggemann, Comparison of Cooling System Designs for an Exhaust Heat Recovery System Using an Organic Rankine Cycle on a Heavy Duty Truck. *Energies* **2016**, *9*, 928. [CrossRef]

58. Karvountzis-Kontakiotis, A.; Pesiridis, A.; Zhao, H.; Alshammari, F.; Franchetti, B.; Pesmazoglou, I.; Tocci, L. *Effect of an ORC Waste Heat Recovery System on Diesel Engine Fuel Economy for Off-Highway Vehicles*; SAE International: Warrendale, PA, USA, 2017.

59. Jadhao, J.S.; Thombare, D.G. Review on Exhaust Gas Heat Recovery for I.C Engine. *Int. J. Eng. Innov. Technol.* **2013**, *2*, 93–100.

60. Hountalas, D.T.; Katsanos, C.O.; Lamaris, V.T. *Recovering Energy from the Diesel Engine Exhaust Using Mechanical and Electrical Turbocompounding*; SAE International: Warrendale, PA, USA, 2007.

61. He, S.; Chang, H.; Zhang, X.; Shu, S.; Duan, C. Working fluid selection for an Organic Rankine Cycle utilizing high and low temperature energy of an LNG engine. *Appl. Therm. Eng.* **2015**, *90*, 579–589. [CrossRef]

62. Noora, A.M.; Putehc, R.C.; Martinez-Botasd, R.; Rajooa, S.; Romagnolie, A.; Basheera, U.M.; Sallehb, S.H.S.; Saha, M.H.M. Technologies for Waste Heat Energy Recovery from Internal Combustion Engine: A Review. In Proceedings of the International Conference on New Trends in Multidisciplinary Research & Practice, Istanbul, Turkey, 4–5 November 2015.

63. Hsieh, J.-C.; Fu, B.-R.; Wang, T.-W.; Cheng, Y.; Lee, Y.-R.; Chang, J.-C. Design and preliminary results of a 20-kW transcritical organic Rankine cycle with a screw expander for low-grade waste heat recovery. *Appl. Therm. Eng.* **2017**, *110*, 1120–1127. [CrossRef]

64. Shu, G.; Wang, X.; Tian, H. Theoretical analysis and comparison of rankine cycle and different organic rankine cycles as waste heat recovery system for a large gaseous fuel internal combustion engine. *Appl. Therm. Eng.* **2016**, *108*, 525–537. [CrossRef]

65. Reck, M.; Randolf, D. *An Organic Rankine Cycle Engine for a 25-Passenger Bus*; SAE International: Warrendale, PA, USA, 1973.

66. Markides, O.A.; Markides, C.N. Thermo-Economic and Heat Transfer Optimization of Working-Fluid Mixtures in a Low-Temperature Organic Rankine Cycle System. *Energies* **2016**, *9*, 448.
67. Brands, M.C.; Werner, J.R.; Hoehne, J.L.; Kramer, S. *Vechicle Testing of Cummins Turbocompound Diesel Engine*; SAE International: Warrendale, PA, USA, 1981.
68. Alshammari, F.; Pesyridis, A.; Karvountzis-Kontakiotis, A.; Franchetti, B.; Pesmazoglou, I. Experimental Study of a Small Scale Organic Rankine Cycle Waste Heat Recovery System for a Heavy Duty Diesel Engine with Focus on the Radial Inflow Turbine Expander Performance. *Appl. Energy* **2018**, *215*, 543–555. [CrossRef]

© 2019 by the authors. Licensee MDPI, Basel, Switzerland. This article is an open access article distributed under the terms and conditions of the Creative Commons Attribution (CC BY) license (http://creativecommons.org/licenses/by/4.0/).

Article

Configuration Optimization and Performance Comparison of STHX-DDB and STHX-SB by A Multi-Objective Genetic Algorithm

Zhe Xu [1,*], Yingqing Guo [1], Haotian Mao [1] and Fuqiang Yang [2]

[1] School of Power and Energy, Northwestern Polytechnical University, Xi'an 710129, China; yqguo@nwpu.edu.cn (Y.G.); alexmao@mail.nwpu.edu.cn (H.M.)

[2] School of Mechanical Engineering, Northwestern Polytechnical University, Xi'an 710072, China; fqyang@nwpu.edu.cn

* Correspondence: zhexu@mail.nwpu.edu.cn

Received: 16 March 2019; Accepted: 7 May 2019; Published: 11 May 2019

Abstract: Based on the thermohydraulic calculation model verified in this study and Non-dominated Sorted Genetic Algorithm-II (NSGA-II), a multi-objective configuration optimization method is proposed, and the performances of shell-and-tube heat exchanger with disc-and-doughnut baffles (STHX-DDB) and shell-and-tube heat exchanger with segmental baffles (STHX-SB) are compared after optimization. The results show that, except in the high range of heat transfer capacity of 16.5–17 kW, the thermohydraulic performance of STHX-DDB is better. Tube bundle diameter, inside tube bundle diameter, number of baffles of STHX-DDB and tube bundle diameter, baffle cut, number of baffles of STHX-SB are chosen as design parameters, and heat transfer capacity maximization and shell-side pressure drop minimization are considered as common optimization objectives. Three optimal configurations are obtained for STHX-DDB and another three are obtained for STHX-SB. The optimal results show that all the six selected optimal configurations are better than the original configurations. For STHX-DDB and STHX-SB, compared with the original configurations, the heat transfer capacity of optimal configurations increases by 6.26% on average and 5.16%, respectively, while the shell-side pressure drop decreases by 44.33% and 19.16% on average, respectively. It indicates that the optimization method is valid and feasible and can provide a significant reference for shell-and-tube heat exchanger design.

Keywords: shell-and-tube heat exchanger; disc-and-doughnut baffles; segmental baffles; multi-objective configuration optimization; genetic algorithm

---

1. Introduction

Shell-and-tube heat exchanger (STHX) is one kind of mechanical device that can exchange heat through a tube wall between two fluids of different temperatures, which are widely used in oil cooling in aero-engines, heating, chemical and food industries, and so on [1]. In the process of heat exchanger design, different configuration parameters can lead to different performances. In order to obtain the better performances, heat exchanger optimization is increasing significantly [2].

In the pursuit of optimized designs, configuration optimization as well as optimization methods application have been analyzed from different points of view. The genetic simulated annealing algorithm [3], biogeography-based algorithm [4], firefly algorithm [5], Jaya algorithm [6] and differential evolution algorithm [7] were used to minimize the cost of the heat exchanger. However, in the actual design process of the heat exchanger, total cost minimization is not the only objective. Many other objectives need to be considered, just like heat transfer capacity maximization, pressure drop minimization, and so on. Thus, the results of research that can consider more than one objective simultaneously could be more meaningful.

In the research of Sanaye et al. [8], the Bell-Delaware procedure and ε-NTU method were used to calculate the shell side heat transfer coefficient and pressure drop of shell-and-tube heat exchanger with segmental baffles (STHX-SB), and the fast and elitist non-dominated sorted genetic algorithm was utilized to solve the optimization problem, which was to maximize the heat transfer coefficient and minimize pressure drop simultaneously, and obtain the Pareto optimal solutions. A method of a multi-objective genetic algorithm combined with numerical simulation was used by Wen et al. [9] and Wang et al. [10,11] to optimize the configuration of STHXs with helical baffles, and the Pareto optimal points of multi-objective optimization were obtained. In addition, this method was also used in the configuration optimization of a spiral-wound heat exchanger by Wang et al. [12]. Some novel theories were also used to design, calculate or optimize heat exchangers. Both Amidpour et al. [13] and Mirzaei et al. [14] applied the constructal theory combined with a genetic algorithm for STHX optimization. An optimization strategy that combined the entransy theory, numerical simulation and a genetic algorithm was proposed by Gu et al. [15] to minimize the entransy dissipation thermal resistance of STHX with helical baffles. Similarly, Chahartaghi et al. [16] used the entransy dissipation theory and Non-dominated Sorted Genetic Algorithm-II (NSGA-II) to minimize the entransy dissipation numbers separately caused by thermal conduction and fluid friction for STHX-SB. In addition, some other novel optimization algorithms were also proposed to optimize STHXs, just like the modified version of teaching-learning-based optimization [17], particle swarm optimization [18], Taguchi method [19], bat algorithm [20], and multi-objective heat transfer search algorithm [21]. In the investigation of Rao et al. [22], an artificial immune system is used with a generalized regression neural network to solve the optimization problem of heat transfer and overall pressure drop for a STHX-SB. Saldanha et al. [23] used Predator–Prey, Multi-objective Particle Swarm Optimization, and NSGA-II evolutionary algorithms to tackle the optimization problem of STHX and chose the best evolutionary algorithms by the Preference Ranking Organization Method for the Enrichment Evaluations decision-making method.

This research had successfully used theoretical approaches or numerical simulation to get the relationships between some configuration parameters and heat exchanger performances, and the corresponding configuration parameters were optimized based on some multi-objective optimization algorithms. Unfortunately, there was only one kind of heat exchanger optimized in each research, and the heat exchanger mentioned was not compared with some different kinds of heat exchangers which are commonly used. Thus, it is worth mentioning that Wang et al. [24] compared heat transfer and pressure drop performances among the STHXs separately with segmental baffles, continuous helical baffles and staggered baffles, and used the multi-objective optimization using a genetic algorithm to obtain the optimal Pareto front of STHX with staggered baffles. However, the comparison was only between the heat exchangers before optimization. The comparison of heat exchangers after optimization under the same condition could reveal their respective advantages and disadvantages after their corresponding configuration parameters were optimal.

In this paper, the configuration parameters of tube bundle diameter, inside tube bundle diameter, number of baffles of shell-and-tube heat exchanger with disc-and-doughnut baffles (STHX-DDB) and the configuration parameters of tube bundle diameter, baffle cut, number of baffles of STHX-SB are analyzed for heat transfer capacity and shell-side pressure drop based on thermohydraulic calculation of the Slipcevie method and Bell-Delaware method [25–29]. In addition, a multi-objective optimization method based on NSGA-II is utilized to optimize these two kinds of heat exchangers and compare the performances of them after optimization.

2. Physical Model and Thermohydraulic Calculation Model

2.1. Physical Model and Configuration Parameters

Both STHX-DDB and STHX-SB primarily consist of heat transfer tubes, baffles, end plates, spacer tubes and shell. The spacer tubes are strung on four heat transfer tubes to locate the positions of baffles. All the baffles are uniformly distributed between two end plates. The differences between disc-and-doughnut baffles and segmental baffles result in the different fluid flow styles of shell-side between STHX-DDB and STHX-SB, as shown in Figure 1. The main configuration parameters of STHX-DDB and STHX-SB are shown in Table 1. The material of STHXs is an aluminium alloy.

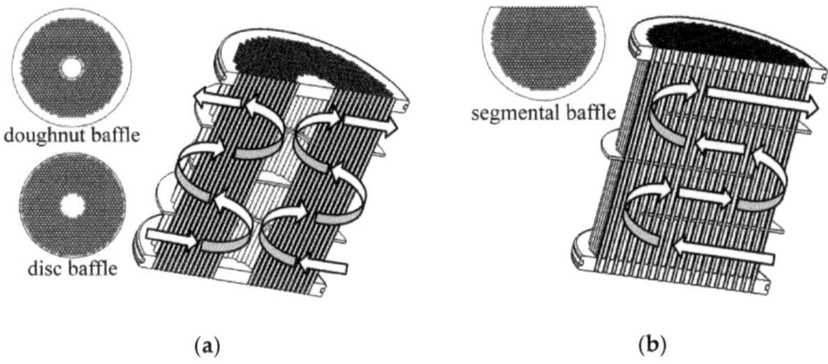

Figure 1. Differences between shell-and-tube heat exchanger with disc-and-doughnut baffles (STHX-DDB) and shell-and-tube heat exchanger with segmental baffles (STHX-SB): (**a**) STHX-DDB; (**b**) STHX-SB.

Table 1. Configuration parameters of STHX-DDB and STHX-SB.

STHX-DDB	Value	STHX-SB	Value
Tube bundle diameter (mm)	100	Tube bundle diameter (mm)	100
Inside tube bundle diameter (mm)	20	Baffle cut	0.25
Inside shell diameter (mm)	115	Inside shell diameter (mm)	115
Tube pitch (mm)	3	Tube pitch (mm)	3
Outside tube diameter (mm)	2.36	Outside tube diameter (mm)	2.36
Inside tube diameter (mm)	1.75	Inside tube diameter (mm)	1.75
Tube length (mm)	130	Tube length (mm)	130
Number of baffles	3	Number of baffles	3
Baffle thickness (mm)	1.5	Baffle thickness (mm)	1.5
Number of end plates	2	Number of end plates	2
End plate thickness (mm)	8	End plate thickness (mm)	8
Number of shell passes	1	Number of shell passes	1
Number of tube passes	2	Number of tube passes	2
Inlet/Outlet diameter of shell/tube-side (mm)	20	Inlet/Outlet diameter of shell/tube-side (mm)	20

To simplify calculations, assumptions are demonstrated as follows: (1) the leakages between baffles and tubes and those between baffles and shell are neglected; (2) the heat transfer between the fluid inside heat exchanger and the environment outside heat exchanger is neglected; (3) both the fluid of shell-side and tube-side are Newtonian and incompressible fluid with constant properties; and (4) the heat exchangers are new and there is no fouling.

2.2. Thermohydraulic Calculation Model

Thermohydraulic calculation of STHXs includes heat transfer capacity calculation and shell-side pressure drop calculation. Basic formulas of heat transfer capacity calculation and shell-side pressure drop calculation are given in Appendices A and B, respectively. The main steps of this thermohydraulic calculation model are shown as follows:

1. Input configuration parameters of STHX as shown in Table 1, and input experiment conditions including inlet temperature/flow rate of tube/shell-side.
2. Assume one heat transfer capacity for STHX.
3. Calculate the mean temperature of tube wall, the physical property parameters (such as fluid density, kinematic viscosity, etc.) and mean temperature of shell/tube-side under the condition of the assumptive heat transfer capacity.
4. Calculate the heat transfer coefficients of shell-side and tube-side.
5. Calculate total heat transfer coefficient, number of transfer units, heat transfer efficiency and heat transfer capacity.
6. Set the assumptive heat transfer capacity equal to the calculated heat transfer capacity and go to step 3 until the deviation between the calculated heat transfer capacity and the assumptive heat transfer capacity is within acceptable limits.
7. Output heat transfer capacity and finish shell-side pressure drop calculation.

The flowchart of thermohydraulic calculation model is shown in Figure 2.

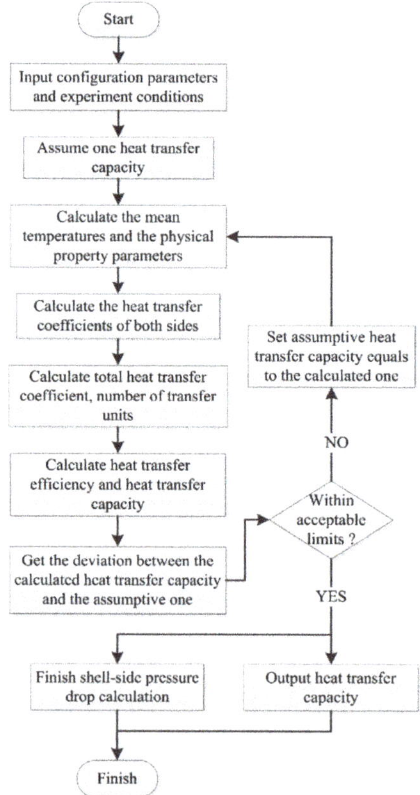

Figure 2. Flowchart of the thermohydraulic calculation model.

3. Optimization Method

3.1. Multi-Objective Genetic Algorithm

It is quite difficult to get the absolute optimal solution of a multi-objective problem because of the conflict of objectives. However, a Pareto optimal front, which consists of a series of solutions, can be obtained through NSGA-II [30]. NSGA-II incorporates elitism [31], and any solution of the Pareto optimal front is generated at the price of other lower priority performances. The ranking scheme of this method adopts the non-dominated sorting method, which is faster than the traditional Pareto ranking method. The constraint handling method also uses a non-dominance principle as the objective, thus penalty functions and Lagrange multipliers are not needed, which guarantees that the feasible solutions are always ranked higher than the unfeasible solutions [32]. A genetic algorithm is a stochastic global searching probabilistic method, which emulates the genetic mechanism of Darwinian evolution and the process of natural selection. This method contains the main steps of selection, crossover and mutation and has the characteristics of self-adaptive, parallel, random [33].

The main steps of the optimization method are shown as follows:

1. Design parameters (such as tube bundle diameter, baffle cut, number of baffles, etc.) and constraints of them; optimization objectives (such as heat transfer capacity maximization, shell-side pressure drop minimization, etc.) are initialized.
2. NSGA-II parameters (such as population size, generation number, etc.) are initialized.
3. A random population is initialized based on population size, number and constrains of design parameters.
4. Calculate objective function values for each chromosome of the population.
5. Each chromosome is sorted based on non-domination and crowding distance. It should be noticed that the crowding distance is compared only if the rank for both chromosomes are the same.
6. The chromosomes suitable for reproduction are selected as parents of the next generation based on tournament algorithm.
7. Children are generated through crossover and mutation, and the objective function values of them are calculated.
8. Parents and children are combined, and each chromosome is sorted based on non-domination and crowding distance.
9. Extract the new generation based on ranking.
10. The above procedures are repeated from step 6 until the convergence.

Details of crossover and mutation are referred in [34]. The flowchart of the optimization method is shown in Figure 3.

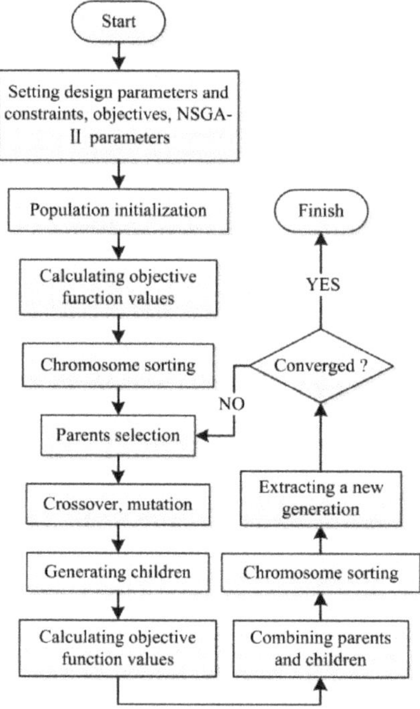

Figure 3. Flowchart of the optimization method.

3.2. Objective Functions and Design Parameters

In this case, in consideration of the thermohydraulic performances of STHX-DDB and STHX-SB, the heat transfer capacity Q and shell-side pressure drop ΔP are selected as objective functions. For STHX-DDB, tube bundle diameter D_o, inside tube bundle diameter D_i and number of baffles N_b are regarded as design parameters, and the multi-objective optimization problem can be formulated as Equation (1):

$$\begin{cases} \text{Min} & -Q, \Delta P \\ \text{S.t.} & 80 \text{ mm} \leq D_o \leq 113 \text{ mm} \\ & 10 \text{ mm} \leq D_i \leq 70 \text{ mm} \\ & N_b = 2, 3, 4, 5, 6 \end{cases} \tag{1}$$

For STHX-SB, tube bundle diameter D_o, baffle cut P and number of baffles N_b are regarded as design parameters, and the multi-objective optimization problem can be formulated as Equation (2):

$$\begin{cases} \text{Min} & -Q, \Delta P \\ \text{S.t.} & 80 \text{ mm} \leq D_o \leq 113 \text{ mm} \\ & 0.15 \leq P \leq 0.45 \\ & N_b = 2, 3, 4, 5, 6 \end{cases} \tag{2}$$

Except for the design parameters, the other configuration parameters of STHX-DDB and STHX-SB are shown in Table 1. The working conditions are shown in Table 2.

Table 2. Experiment conditions for STHX-DDB and STHX-SB.

Experiment Condition	Value
Inlet flow rate of tube-side (L/min)	20
Inlet flow rate of shell-side (L/min)	40
Inlet temperature of tube-side (K)	323
Inlet temperature of shell-side (K)	383

4. Results and Discussion

4.1. Model Validation

In order to validate the accuracy of thermohydraulic calculation model, calculation results are compared with experimental data. In this case, the tube-side fluids of these two heat exchangers, which are heated, are both fuel, and the shell-side fluids, which are cooled, are both oil. The experiment conditions and the testing equipment for STHX-DDB and STHX-SB are given in Table 2 and Figure 4, respectively, and the comparison results are shown in Figure 5.

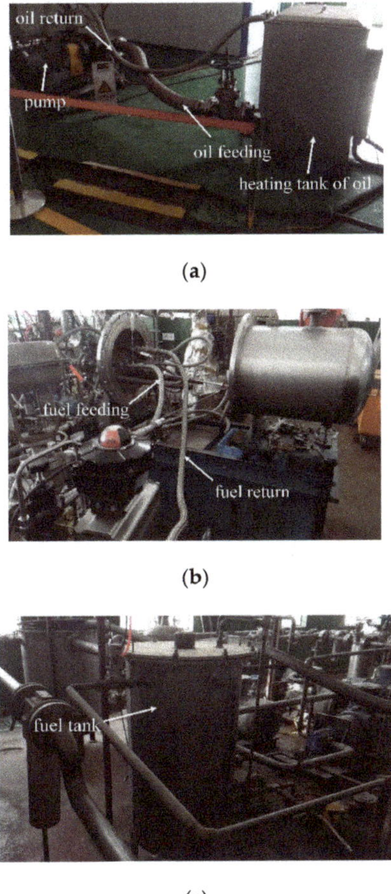

Figure 4. Testing equipment: (**a**) testing equipment of the oil side; (**b**) one part of testing equipment of the fuel side; (**c**) the other part of testing equipment of the fuel side.

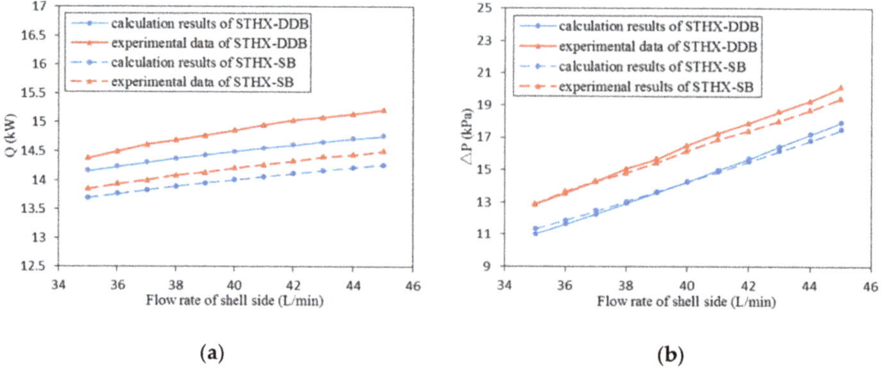

Figure 5. Comparison between calculation results and experimental data: (**a**) heat transfer capacity; (**b**) shell-side pressure drop.

The results illustrate that both the heat transfer capacity and the shell-side pressure drop obtained from calculation model are in agreement with the experimental results. The errors of the heat transfer capacity of STHX-DDB and STHX-SB are 1.5–3.0% and 1.1–1.7% with the average deviations of 2.4% and 1.4%, respectively, and those of shell-side pressure drop are 10.8–14.4% and 10.0–13.0% with the average deviations of 13.0% and 11.6%, respectively. Hence, it can be concluded that the thermohydraulic calculation model is reliable and the errors are acceptable. The differences may be caused by simplification of physical model, deviation of formula itself, and unavoidable measurement error in experiment.

4.2. Design Parameters Effects of STHX-DDB

4.2.1. Effects of Tube Bundle Diameter

The effects of tube bundle diameter D_o on heat transfer capacity Q and shell-side pressure drop ΔP of STHX-DDB are represented in Figure 6 while inside tube bundle diameter and number of baffles are 20 mm and 3, respectively. As depicted, while tube bundle diameter increases from 80 mm to 113 mm, heat transfer capacity increases by 11.47–16.19 kW, and shell-side pressure drop drops slowly from 14.76 kPa first, reaching its lowest point of 14.27 kPa when tube bundle diameter is 102 mm, and then rises more and more dramatically to 19.02 kPa. As shown in Figure 7, as tube bundle diameter increases, the number of tubes increases. Thus, heat transfer area increases, which enhances heat transfer capacity. In addition, the growth of tube bundle diameter means that the flow area between baffles increases and the flow area between disc baffles and shell decreases, which results in the decrease of the pressure drop between baffles and the increase of the pressure drop between disc baffles and shell. Moreover, the pressure drop between disc baffles and shell grows increasingly significantly after tube bundle diameter reaches 102 mm. Thus, shell-side pressure drop falls increasingly slowly first and then grows increasingly sharply after tube bundle diameter reaches 102 mm.

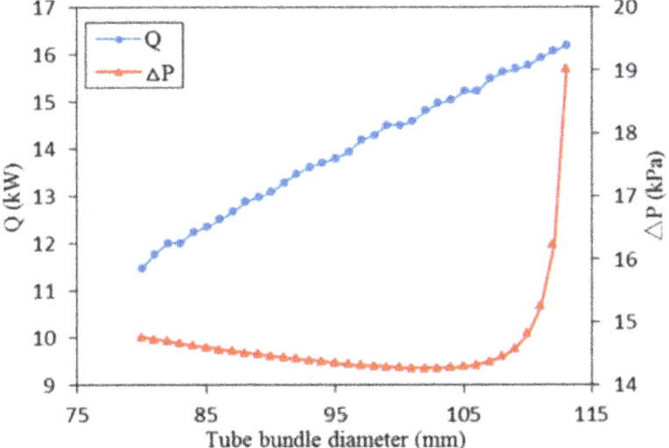

Figure 6. Heat transfer capacity and shell-side pressure drop separately versus tube bundle diameter.

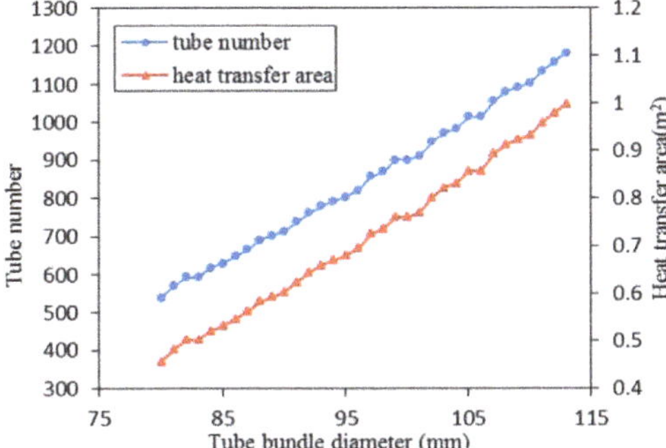

Figure 7. Tube number and heat transfer area separately versus tube bundle diameter.

4.2.2. Effects of Inside Tube Bundle Diameter

The effects of inside tube bundle diameter D_i on heat transfer capacity Q and shell-side pressure drop ΔP of STHX-DDB are represented in Figure 8 while tube bundle diameter and number of baffles are 100 mm and 3, respectively. As depicted, while inside tube bundle diameter increases from 10 mm to 70 mm, heat transfer capacity decreases by 14.82–9.80 kW, and shell-side pressure drop drops significantly from 154.42 kPa first, reaching the point of 5.96 kPa when inside tube bundle diameter is 32 mm, and then falls slightly to 4.05 kPa. As shown in Figure 9, the increase of inside tube bundle diameter results in the decrease of number of tubes and the growth of flow area of the central hole of doughnut baffles. Thus, both heat transfer capacity and shell-side pressure drop decrease. As the impact on shell-side pressure drop is more serious before the point of 5.96 kPa than after, the figure decreases more seriously before this point than after.

4.2.3. Effects of Number of Baffles

The effects of number of baffles N_b on heat transfer capacity Q and shell-side pressure drop ΔP of STHX-DDB are represented in Figure 10 while tube bundle diameter and inside tube bundle diameter are 100 mm and 20 mm, respectively. As depicted, while number of baffles increases from 2 to 6, heat transfer capacity and shell-side pressure drop increase by 14.21–14.95 kW and 8.63–25.60 kPa, respectively. As number of baffles increases, the distance between baffles decreases, which leads to the increase of fluid velocity between baffles. Thus, both heat transfer capacity and shell-side pressure drop increase. In addition, as number of baffles grows, disc baffle number and doughnut baffle number increase alternately, which results in an unsmooth figure of shell-side pressure drop.

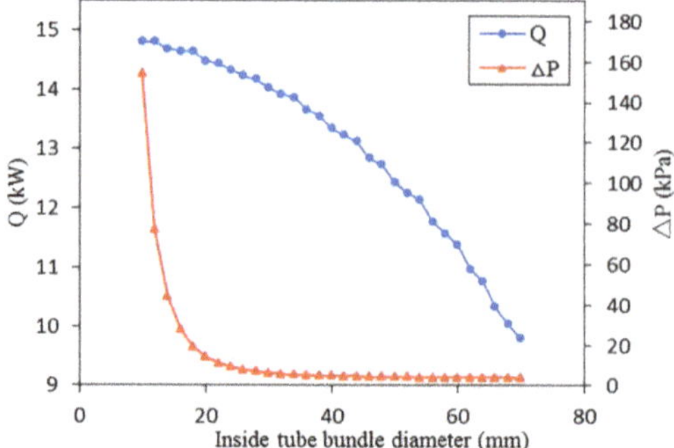

Figure 8. Heat transfer capacity and shell-side pressure drop separately versus inside tube bundle diameter.

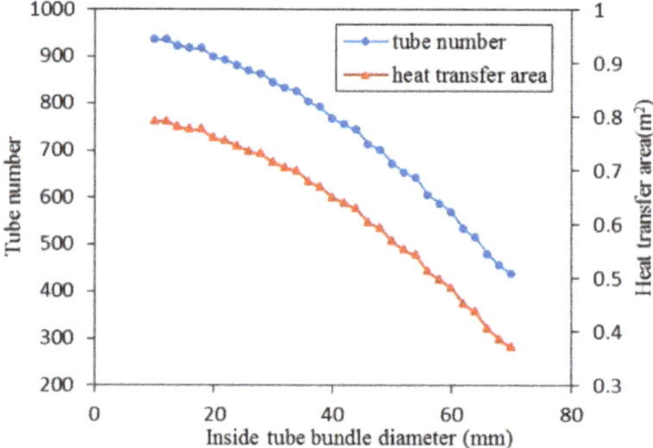

Figure 9. Tube number and heat transfer area separately versus inside tube bundle diameter.

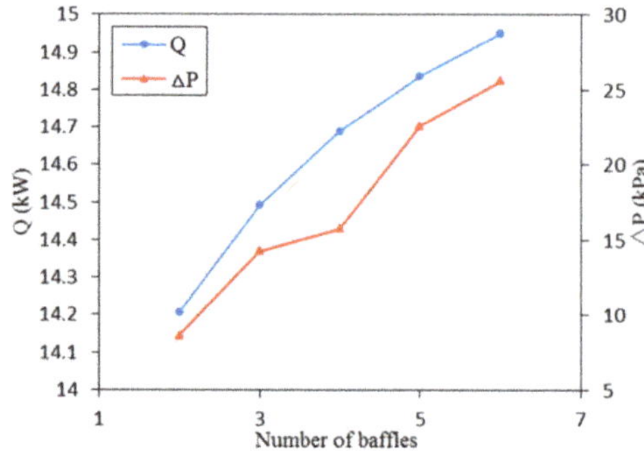

Figure 10. Heat transfer capacity and shell-side pressure drop separately versus number of baffles.

4.3. Design Parameter Effects of STHX-SB

4.3.1. Effects of Tube Bundle Diameter

The effects of tube bundle diameter D_o on heat transfer capacity Q and shell-side pressure drop ΔP of STHX-SB are represented in Figure 11 while baffle cut and number of baffles are 0.25 and 3, respectively. As depicted, while tube bundle diameter increases from 80 mm to 113 mm, heat transfer capacity and shell-side pressure drop increase by 10.16–16.60 kW and 6.59–57.07 kPa, respectively. As shown in Figure 12, as tube bundle diameter grows, the number of tubes increases, which improves heat transfer capacity. At the same time, both the flow areas of cross flow region and window region decrease, which leads to a higher shell-side pressure drop.

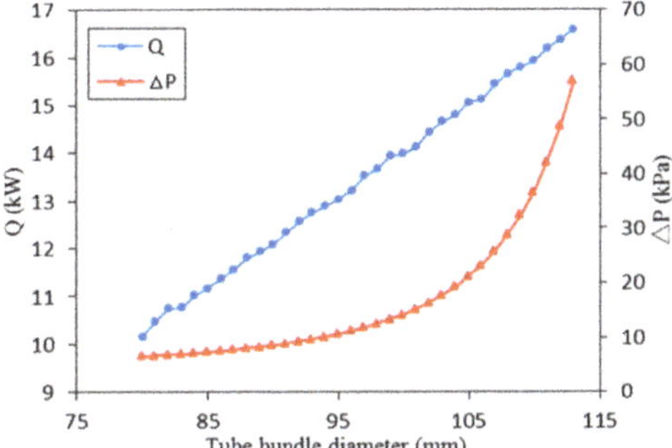

Figure 11. Heat transfer capacity and shell-side pressure drop separately versus tube bundle diameter.

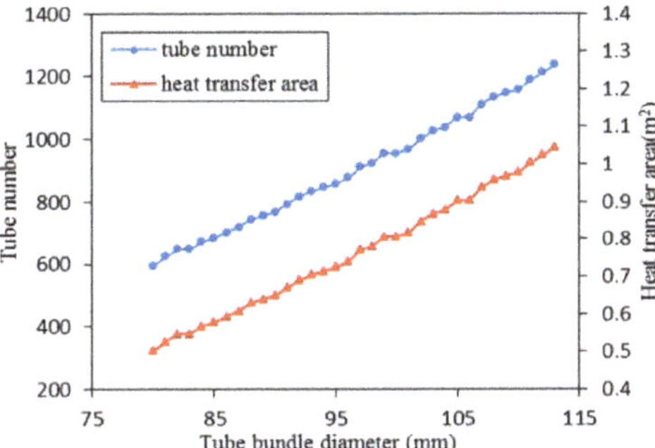

Figure 12. Tube number and heat transfer area separately versus tube bundle diameter.

4.3.2. Effects of Baffle Cut

The effects of baffle cut P on heat transfer capacity Q and shell-side pressure drop ΔP of STHX-SB are represented in Figure 13 while tube bundle diameter and number of baffles are 100 mm and 3, respectively. As depicted, while baffle cut increases from 0.15 to 0.45, heat transfer capacity and shell-side pressure drop decrease by 14.50–12.44 kW and 16.74–9.99 kPa, respectively. The increase of baffle cut means that cross flow decreases and parallel flow increases. As cross flow has a better heat transfer effect and a bigger pressure drop than parallel flow, both heat transfer capacity and shell-side pressure drop decrease.

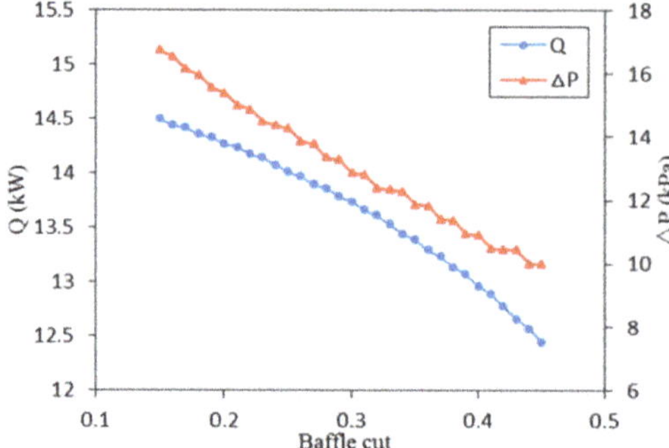

Figure 13. Heat transfer capacity and shell-side pressure drop separately versus baffle cut.

4.3.3. Number of Baffles

The effects of number of baffles N_b on heat transfer capacity Q and shell-side pressure drop ΔP of STHX-SB are represented in Figure 14 while tube bundle diameter and baffle cut are 100 mm and 0.25, respectively. As depicted, while number of baffles increases from 2 to 6, heat transfer capacity and shell-side pressure drop increase by 13.88–14.25 kW and 9.17–43.22 kPa, respectively. As number of baffles increases, turbulence of fluid is strengthened. Thus, both heat transfer capacity and shell-side pressure drop increase.

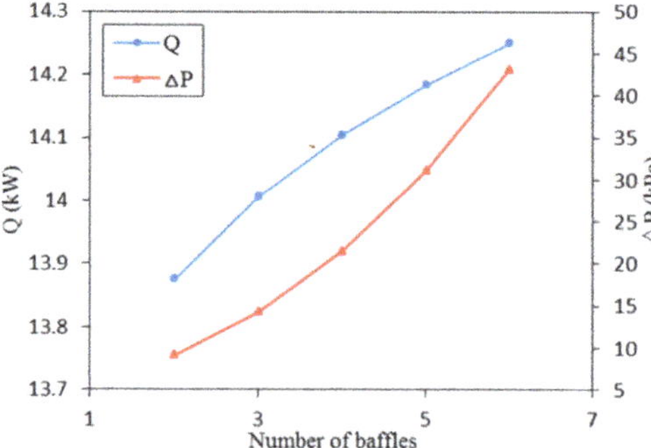

Figure 14. Heat transfer capacity and shell-side pressure drop separately versus number of baffles.

4.4. Multi-Objective Optimization Results

In order to consider the comprehensive thermohydraulic performances of STHX-DDB and STHX-SB, the multi-objective configuration optimization using NSGA-II is conducted. The optimization goals are to maximize heat transfer capacity and to minimize shell-side pressure drop, and two objective functions ($-Q$ and ΔP) are conflicting. Population size, maximum iteration number, analog binary cross distribution index and polynomial mutation distribution index, which are main settings of optimization, are 100, 500, 20 and 20, respectively. Figure 15 shows the Pareto optimal points which are obtained based on the objective functions and the design parameters described in Equations (1) and (2), and the Pareto optimal points in the heat transfer capacity ranges of 7.5–16.5 kW and 16.5–17 kW are shown in Figures 16 and 17, respectively.

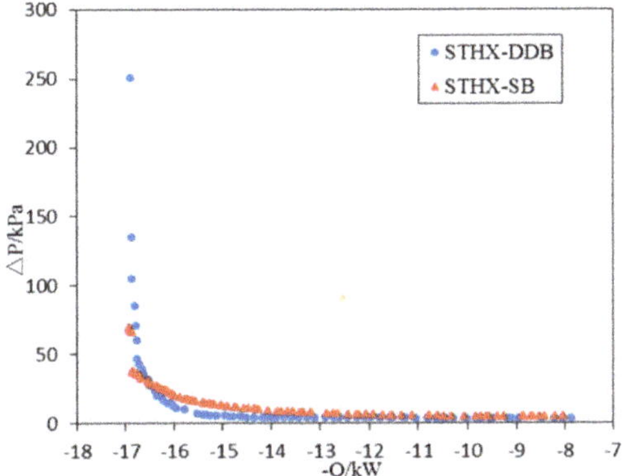

Figure 15. Pareto optimal points.

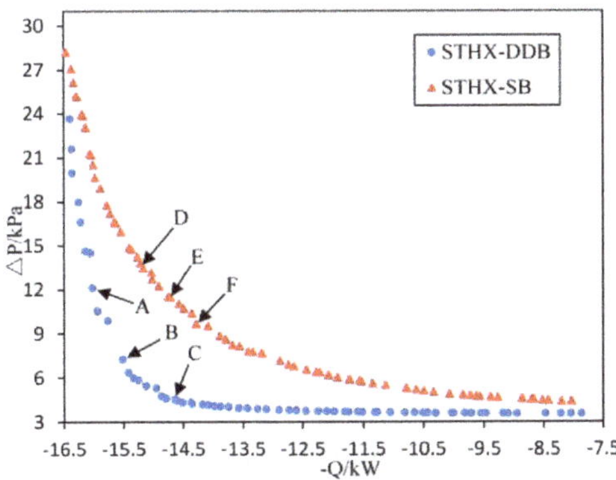

Figure 16. Pareto optimal points for the heat transfer capacity range of 7.5–16.5 kW.

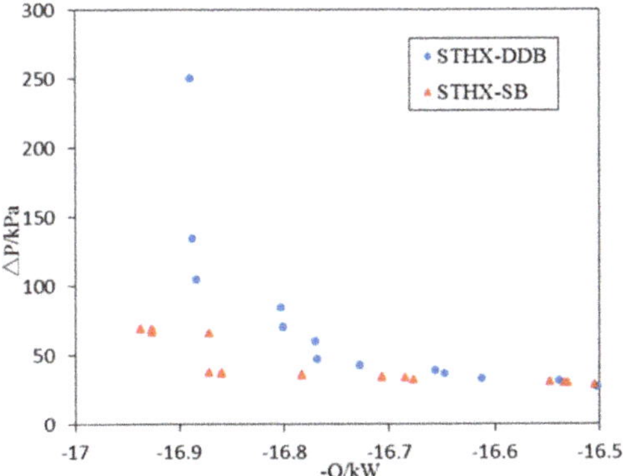

Figure 17. Pareto optimal points for the heat transfer capacity range of 16.5–17 kW.

As shown in Figure 15, for both heat exchangers, shell-side pressure drop increases with the growth of heat transfer capacity. Especially in the range of high heat transfer capacity, shell-side pressure drop increases significantly sharply. As depicted in Figure 16, after optimization, for the same heat transfer capacity, the shell-side pressure drop of STHX-DDB is lower than this value of STHX-SB in the heat transfer capacity range of 7.5–16.5 kW. Thus, STHX-DDB has a better thermohydraulic performance than STHX-SB in this range. As shown in Figure 17, after optimization, for the same heat transfer capacity, the shell-side pressure drop of STHX-DDB is above or almost equal to this value of STHX-SB in the heat transfer capacity range of 16.5–17 kW. Thus, in this high range of heat transfer capacity, the thermohydraulic performance of STHX-SB is better.

Three Pareto optimal solutions for STHX-DDB and three Pareto optimal solutions for STHX-SB are chosen and shown in Figure 16, Tables 3 and 4. For either STHX-DDB or STHX-SB, the selection method of three Pareto optimal solutions follows the following principles:

1. The heat transfer capacity of the first Pareto optimal solution is much higher than the original configuration, while the shell-side pressure drop of it is a little lower than the original configuration.
2. The heat transfer capacity of the third Pareto optimal solution is a little higher than the original configuration, while the shell-side pressure drop of it is much lower than the original configuration.
3. The second Pareto optimal solution is chosen from almost the middle position in the range from the first Pareto optimal solution to the third.

Table 3. Optimization thermohydraulic performances comparison of STHX-DDB.

Parameters	D_o (mm)	D_i (mm)	N_b	Q (kW)	ΔP (kPa)
Original configuration	100	20	3	14.49	14.28
Optimal configuration A	112	23	3	16.02	12.11
Optimal configuration B	111	24	2	15.52	7.24
Optimal configuration C	107	34	2	14.65	4.50

Table 4. Optimization thermohydraulic performances comparison of STHX-SB.

Parameters	D_o (mm)	P	N_b	Q (kW)	ΔP (kPa)
Original configuration	100	0.25	3	14.01	14.25
Optimal configuration D	105	0.21	2	15.18	13.45
Optimal configuration E	103	0.22	2	14.73	11.47
Optimal configuration F	99	0.17	2	14.29	9.64

Compared with the original configuration of STHX-DDB shown in Table 3, the heat transfer capacity values of optimal configuration A, B and C increase separately by 10.56%, 7.11% and 1.10%, and the shell-side pressure drop values of them decrease by 15.20%, 49.30% and 68.49%, respectively. It is clearly depicted that the heat transfer capacity of optimal configurations of STHX-DDB increases by 6.26% on average, while the shell-side pressure drop decreases by 44.33% on average. Compared with the original configuration of STHX-SB shown in Table 4, the heat transfer capacity values of optimal configuration D, E and F increase separately by 8.35%, 5.14% and 2.00%, and the shell-side pressure drop values of them decrease by 5.61%, 19.51% and 32.35%, respectively. It is clearly shown that the heat transfer capacity of optimal configurations of STHX-SB increases by 5.16% on average, while the shell-side pressure drop decreases by 19.16% on average. Based on the comparison and analysis of original and optimal configurations, it indicates that the optimization method is valid and feasible, and the comprehensive thermohydraulic performances of STHX-DDB and STHX-SB can be enhanced by the optimization method.

5. Conclusions

In this paper, a multi-objective configuration optimization method for STHX-DDB and STHX-SB based on NSGA-II is proposed and used. Based on the thermohydraulic calculation model verified in this paper, the effects of design parameters are analyzed. Then, the optimal design parameters (tube bundle diameter, inside tube bundle diameter, number of baffles) for STHX-DDB and the optimal design parameters (tube bundle diameter, baffle cut, number of baffles) for STHX-SB are found separately by a trade-off between the two common objectives of heat transfer capacity maximization and shell-side pressure drop minimization.

The distribution of Pareto optimal points reveals that the two objectives are conflicted, and, for both heat exchangers, shell-side pressure drop increases significantly sharply in the range of high heat transfer capacity, which means that high heat transfer capacity will cost a high shell-side pressure drop. In addition, after optimization, except in the high range of heat transfer capacity of 16.5–17 kW, the shell-side pressure drop of STHX-DDB is lower than this value of STHX-SB when the heat transfer capacity values of them are the same. In other words, under this condition, the thermohydraulic performance of STHX-DDB is better.

Finally, three optimal solutions for STHX-DDB and three optimal solutions for STHX-SB are selected and compared with the original configurations. The results illustrate that all the selected optimal configurations are better than the original configurations. For STHX-DDB, the heat transfer capacity of optimal configurations increases by 6.26% on average compared with the original configuration, while the shell-side pressure drop decreases by 44.33% on average. For STHX-SB, the heat transfer capacity of optimal configurations increases by 5.16% on average compared with the original configuration, while the shell-side pressure drop decreases by 19.16% on average. It indicates that the optimization method is feasible and valid and can provide a significant reference for STHX design.

Author Contributions: Conceptualization, Z.X.; Investigation, Z.X. and H.M.; Methodology, Z.X.; Project administration, Y.G.; Resources, Z.X.; Software, Z.X.; Supervision, Y.G.; Validation, Z.X.; Visualization, Z.X.; Writing—original draft, Z.X.; Writing—review and editing, Z.X., Y.G. and F.Y.

Funding: This research received no external funding.

Conflicts of Interest: The authors declare no conflict of interest.

Nomenclature

Latin letters

A	heat transfer area	m^2
C^*	heat capacity ratio	-
C_p	specific heat capacity	J/(kg·K)
d_e	equivalent diameter	m
d_i	inside tube diameter	m
d_m	average tube diameter	m
d_o	outside tube diameter	m
f	friction coefficient	-
K	total heat transfer coefficient	W/(m²·K)
L	tube length	m
l_b	space between baffles	m
n	total number of tube rows	-
N_B	number of baffles	-
N_{B1}	number of disc baffles	-
N_{B2}	number of doughnut baffles	-
N_c	umber of tubes in cross flow region	-
N_{cw}	number of tubes in window region	-
NTU	number of transfer units	-
Pr	Prandtl number	-
P_t	tube pitch	m
Q	heat transfer capacity	W
q_m	mass flow rate	kg/s
Re	Reynolds number	-
$r_{s,i}$	fouling resistance inside tube	m^2·K/W

$r_{s,o}$	fouling resistance outside tube	$\mathrm{m^2 \cdot K/W}$
S	circulation area	$\mathrm{m^2}$
t	temperature	K
t_1	inlet temperature of hot side	K
t_2	inlet temperature of cold side	K
u	fluid velocity	m/s
W	heat capacity	W/K
Greek letters		
α	heat transfer coefficient	$\mathrm{W/(m^2 \cdot K)}$
δ	thickness	m
ΔP	shell-side pressure drop	Pa
ε	resistance coefficient	-
η	heat transfer efficiency	-
λ	heat conductivity	$\mathrm{W/(m \cdot K)}$
μ	fluid viscosity	$\mathrm{Pa \cdot s}$
ρ	fluid density	$\mathrm{kg/m^3}$
Subscripts		
i	tube-side	
o	shell-side	
w	tube wall	
N	outlet/inlet	
B	between baffles	
$s1$	between disc baffles and shell	
$s2$	central hole of doughnut baffles	
bk	cross flow region	
wk	window region	

Appendix A

Basic formulas of heat transfer capacity calculation. The basic formula of heat transfer capacity calculation of heat exchanger is given by Equation (A1) [29]:

$$Q = \eta W_{\min}(t_1 - t_2). \tag{A1}$$

As the number of shell passes and tube passes are 1 and 2, respectively, η is given by Equations (A2)–(A6) [29,35]:

$$\eta = \frac{2}{1 + C^* + \sqrt{1 + C^{*2}}\left(1 + e^{-\mathrm{NTU}\sqrt{1+C^{*2}}}\right)/\left(1 - e^{-\mathrm{NTU}\sqrt{1+C^{*2}}}\right)}, \tag{A2}$$

$$C^* = W_{\min}/W_{\max}, \tag{A3}$$

$$\mathrm{NTU} = \frac{KA}{W_{\min}}, \tag{A4}$$

$$\frac{1}{K} = \frac{1}{\alpha_i}\left(\frac{d_o}{d_i}\right) + r_{s,i}\left(\frac{d_o}{d_i}\right) + \frac{\delta_w}{\lambda_w}\left(\frac{d_o}{d_m}\right) + r_{s,o} + \frac{1}{\alpha_o}, \tag{A5}$$

$$d_m = \frac{d_o - d_i}{\ln(d_o/d_i)}. \tag{A6}$$

In general, the thermal-conduction resistance of a metal tube wall is much smaller than the thermal resistance of fluids' heat convection, and, for the new heat exchanger, the fouling resistance can be neglected. In addition, the tube wall thickness used in this case is only 0.305 mm, which is too small to be ignored. Thus, Equation (A5) can be simplified to Equation (A7) [29]:

$$K = \frac{\alpha_o \alpha_i}{\alpha_o + \alpha_i}. \tag{A7}$$

For STHXs, α_i is given by Equations (A8)–(A10) [26]:

$$\alpha_i = 0.027 \frac{\lambda_i}{d_i} Re_i^{0.8} \Pr_i^{1/3} \left(\frac{\mu_i}{\mu_{iw}}\right)^{0.14} \tag{A8}$$

$(Re_i > 10{,}000, \; 0.7 < Pr_i < 16{,}700, \; \frac{L}{d_i} \geq 60)$,

$$\alpha_i = 0.027 \left(1 - \frac{6 \times 10^5}{Re_i^{1.8}}\right) \frac{\lambda_i}{d_i} Re_i^{0.8} \Pr_i^{1/3} \left(\frac{\mu_i}{\mu_{iw}}\right)^{0.14} \tag{A9}$$

$(2300 \leq Re_i \leq 10{,}000, \; 0.7 < Pr_i < 16{,}700, \; \frac{L}{d_i} \geq 60)$,

$$\alpha_i = 1.86 \frac{\lambda_i}{d_i} \left(Re_i \Pr_i \frac{d_i}{L}\right)^{1/3} \left(\frac{\mu_i}{\mu_{iw}}\right)^{0.14} \tag{A10}$$

$(Re_i < 2300, \; 0.6 < Pr_i < 6700, \; Re_i Pr_i \frac{L}{d_i} \geq 100)$.

In Equations (A8)–(A10), μ_{iw} is the viscosity of tube-side fluid under the temperature of tube wall.

Because of the different shell-side structures between STHX-DDB and STHX-SB, α_o calculation formulas are different. For STHX-DDB, the Slipcevie method is used, and α_o is given by Equations (A11) and (A12) [26]:

$$\alpha_o = \sum_{j=1}^{n} \frac{A(j)}{A} \alpha_o(j), \tag{A11}$$

$$\alpha_o(j) = 0.33 \frac{\lambda_o}{d_o} Re_o^{0.6}(j) \Pr_o^{0.33} \left(\frac{t_o}{t_w}\right)^{0.14}. \tag{A12}$$

In Equations (A11) and (A12), $A(j)$, $\alpha_o(j)$ and $Re_o(j)$ represent heat transfer area, heat transfer coefficient and Reynolds number of the fluid over the tube row of j, respectively.

For STHX-SB, the Bell–Delaware method is used and α_o is given by Equation (A13) [26–29]:

$$\alpha_o = \alpha_{id} J_c J_l J_b J_s J_r. \tag{A13}$$

In Equation (A13), α_{id} is the ideal heat transfer coefficient for pure cross flow over a tube bundle. J_c, J_l, J_b, J_s and J_r represent correction factors for baffle cut, baffle leakage, bundle bypassing, variable baffle spacing at the inlet and outlet of shell, and adverse temperature gradient in laminar flow, respectively.

Re and Pr used in the calculation above can be calculated by Equations (A14)–(A16) [29]:

$$Re = \frac{\rho u d_e}{\mu}, \tag{A14}$$

$$\Pr = \frac{\mu C_p}{\lambda}, \tag{A15}$$

$$u = \frac{q_m}{\rho S}. \tag{A16}$$

Appendix B

Basic formulas of shell-side pressure drop calculation. Because of the different shell-side structures between STHX-DDB and STHX-SB, formulas of ΔP calculation are different.

For STHX-DDB, ΔP is given by Equations (A17)–(A24) [36]:

$$\Delta P = \Delta P_N + \Delta P_B + \Delta P_{s1} + \Delta P_{s2}, \tag{A17}$$

$$\Delta P_N = 1.5 \frac{\rho_0 u_N^2}{2}, \tag{A18}$$

$$\Delta P_B = \varepsilon_B (N_B + 1) \frac{\rho_0 u_B^2}{2}, \tag{A19}$$

$$\Delta P_{s1} = N_{B1} \varepsilon_{s1} \frac{\rho_0 u_{s1}^2}{2}, \tag{A20}$$

$$\Delta P_{s2} = N_{B2} \varepsilon_{s2} \frac{\rho_0 u_{s2}^2}{2}, \tag{A21}$$

$$\varepsilon_B = \begin{cases} \dfrac{60}{\frac{L-d_o}{d_o} Re_B} & \left(Re_B < \frac{42.3 d_o}{L-d_o} \right) \\ \dfrac{3}{\left(\frac{L-d_o}{d_o} Re_B \right)^{0.2}} & \left(Re_B \geq \frac{42.3 d_o}{L-d_o} \right) \end{cases}, \tag{A22}$$

$$\varepsilon_{s1} = 2.2 + 286 Re_{s1}^{-0.845}, \tag{A23}$$

$$\varepsilon_{s2} = 2.2 + 286 Re_{s2}^{-0.845}. \tag{A24}$$

For STHX-SB, the Bell-Delaware method is used and ΔP is given by Equations (A25)–(A28) [26–28].

$$\Delta P = 2\Delta P_{bk} R_b \left(1 + \frac{N_{cw}}{N_c} \right) R_s + [(N_B - 1)\Delta P_{bk} R_b + N_B \Delta P_{wk}] R_l + \Delta P_N, \tag{A25}$$

$$\Delta P_{bk} = 4 f_o N_c \frac{\rho_0 u_{bk}^2}{2} \left(\frac{\mu_o}{\mu_{ow}} \right)^{-0.14}, \tag{A26}$$

$$f_o = \begin{cases} 47.1 Re_{bk}^{-0.965} & (Re_{bk} < 100) \\ 13.0 Re_{bk}^{-0.685} & (100 \leq Re_{bk} \leq 300) \\ 3.2 Re_{bk}^{-0.44} & (300 < Re_{bk} \leq 1000) \\ 0.505 Re_{bk}^{-0.176} & (Re_{bk} > 1000) \end{cases}, \tag{A27}$$

$$\Delta P_{wk} = \begin{cases} (2 + 0.6 N_{cw}) \frac{\rho_0 u_{wk}^2}{2} & (Re_{wk} \geq 100) \\ 26 \mu_o u_{wk} \left(\frac{N_{cw}}{p_t - d_o} + \frac{l_b}{d_{ew}^2} \right) + 2 \left(\frac{\rho_0 u_{wk}^2}{2} \right) & (Re_{wk} < 100) \end{cases}. \tag{A28}$$

In Equations (A25)–(A28), R_b, R_l, R_s represent correction factors for bundle bypassing, baffle leakage, and variable baffle spacing at the inlet and outlet of shell, respectively. μ_{ow} is the viscosity of shell-side fluid under the temperature of tube wall. d_{ew} is the equivalent diameter of tube bundle in the window region.

References

1. Bayram, H.; Sevilgen, G. Numerical Investigation of the Effect of Variable Baffle Spacing on the Thermal Performance of a Shell and Tube Heat Exchanger. *Energies* **2017**, *10*, 1156. [CrossRef]
2. Wang, Y.; Huai, X. Heat Transfer and Entropy Generation Analysis of an Intermediate Heat Exchanger in ADS. *J. Therm. Sci.* **2018**, *27*, 175–183. [CrossRef]
3. Xiao, W.; Wang, K.; Jiang, X.; He, G. Optimization of a shell-and-tube heat exchanger based on a genetic simulated annealing algorithm. *J. Tsinghua Univ. (Sci. Technol.)* **2016**, *56*, 728–734.
4. Hadidi, A.; Nazari, A. Design and economic optimization of shell-and-tube heat exchangers using biogeography-based (BBO) algorithm. *Appl. Therm. Eng.* **2013**, *51*, 1263–1272. [CrossRef]
5. Mohanty, D.K. Application of firefly algorithm for design optimization of a shell and tube heat exchanger from economic point of view. *Int. J. Therm. Sci.* **2016**, *102*, 228–238. [CrossRef]
6. Rao, R.V.; Saroj, A. Economic optimization of shell-and-tube heat exchanger using Jaya algorithm with maintenance consideration. *Appl. Therm. Eng.* **2017**, *116*, 473–487. [CrossRef]

7. Singh, P.; Pant, M. Design Optimization of Shell and Tube Heat Exchanger Using Differential Evolution Algorithm. In *Proceedings of the Third International Conference on Soft Computing for Problem Solving*; Springer: New Delhi, India, 2014.

8. Sanaye, S.; Hajabdollahi, H. Multi-objective optimization of shell and tube heat exchangers. *Appl. Therm. Eng.* **2010**, *30*, 1937–1945. [CrossRef]

9. Wen, J.; Gu, X.; Wang, M.; Wang, S.; Tu, J. Numerical investigation on the multi-objective optimization of a shell-and-tube heat exchanger with helical baffles. *Int. Commun. Heat Mass* **2017**, *89*, 91–97. [CrossRef]

10. Wang, S.; Xiao, J.; Wang, J.; Jian, G.; Wen, J.; Zhang, Z. Configuration optimization of shell-and-tube heat exchangers with helical baffles using multi-objective genetic algorithm based on fluid-structure interaction. *Int. Commun. Heat Mass* **2017**, *85*, 62–69. [CrossRef]

11. Wang, S.; Xiao, J.; Wang, J.; Jian, G.; Wen, J.; Zhang, Z. Application of response surface method and multi-objective genetic algorithm to configuration optimization of shell-and-tube heat exchanger with fold helical baffles. *Appl. Therm. Eng.* **2018**, *129*, 512–520. [CrossRef]

12. Wang, S.; Jian, G.; Xiao, J.; Wen, J.; Zhang, Z.; Tu, J. Fluid-thermal-structural analysis and structural optimization of spiral-wound heat exchanger. *Int. Commun. Heat Mass* **2018**, *95*, 42–52. [CrossRef]

13. Amidpour, M.; Azad, A.V. Economic optimization of shell and tube heat exchanger based on constructal theory. In Proceedings of the ASME 2010 4th International Conference on Energy Sustainability, Phoenix, AZ, USA, 17–22 May 2010.

14. Mirzaei, M.; Hajabdollahi, H.; Fadakar, H. Multi-objective optimization of shell-and-tube heat exchanger by constructal theory. *Appl. Therm. Eng.* **2017**, *125*, 9–19. [CrossRef]

15. Gu, X.; Wang, M.; Liu, Y.; Wang, S. Multi-parameter optimization of shell-and-tube heat exchanger with helical baffles based on entransy theory. *Appl. Therm. Eng.* **2018**, *130*, 804–813.

16. Chahartaghi, M.; Eslami, P.; Naminezhad, A. Effectiveness improvement and optimization of shell-and-tube heat exchanger with entransy method. *Heat Mass Transf.* **2018**, *54*, 3771–3784. [CrossRef]

17. Rao, R.V.; Patel, V. Multi-objective optimization of heat exchangers using a modified teaching-learning-based optimization algorithm. *Appl. Math. Model.* **2013**, *37*, 1147–1162. [CrossRef]

18. Ghanei, A.; Assareh, E.; Biglari, M.; Ghanbarzadeh, A.; Noghrehabadi, A.R. Thermal-economic multi-objective optimization of shell and tube heat exchanger using particle swarm optimization (PSO). *Heat Mass Transf.* **2014**, *50*, 1375–1384. [CrossRef]

19. Etghani, M.M.; Baboli, S.A.H. Numerical investigation and optimization of heat transfer and exergy loss in shell and helical tube heat exchanger. *Appl. Therm. Eng.* **2017**, *121*, 294–301. [CrossRef]

20. Tharakeshwar, T.K.; Seetharamu, K.N.; Prasad, B.D. Multi-objective optimization using bat algorithm for shell and tube heat exchangers. *Appl. Therm. Eng.* **2017**, *110*, 1029–1038. [CrossRef]

21. Raja, B.D.; Jhala, R.L.; Patel, V. Many-objective optimization of shell and tube heat exchanger. *Therm. Sci. Eng. Prog.* **2017**, *2*, 87–101. [CrossRef]

22. Rao, B.B.; Raju, V.R.; Deepak, B. Estimation and optimization of heat transfer and overall pressure drop for a shell and tube heat exchanger. *J. Mech. Sci. Technol.* **2017**, *31*, 375–383. [CrossRef]

23. Saldanha, W.H.; Soares, G.L.; Machado-Coelho, T.M.; dos Santos, E.D.; Ekel, P.I. Choosing the best evolutionary algorithm to optimize the multiobjective shell-and-tube heat exchanger design problem using PROMETHEE. *Appl. Therm. Eng.* **2017**, *127*, 1049–1061. [CrossRef]

24. Wang, X.; Zheng, N.; Liu, Z.; Liu, W. Numerical analysis and optimization study on shell-side performances of a shell and tube heat exchanger with staggered baffles. *Int. J. Heat Mass Transf.* **2018**, *124*, 247–259. [CrossRef]

25. Gu, X.; Zheng, Z.; Xiong, X.; Wang, T.; Luo, Y.; Wang, K. Characteristics of Fluid Flow and Heat Transfer in the Shell Side of the Trapezoidal-like Tilted Baffles Heat Exchanger. *J. Therm. Sci.* **2018**, *27*, 602–610. [CrossRef]

26. Qian, S. *Heat Exchanger Design Manual*; Chemistry Industry Publisher: Beijing, China, 2002; pp. 55–108.

27. Thulukkanam, K. *Heat Exchanger Design Handbook*, 2nd ed.; CRC Press: New York, NY, USA, 2013; pp. 39–115.

28. Shah, R.K.; Sekulic, D.P. *Fundamentals of Heat Exchanger Design*; John Wiley & Sons: New York, NY, USA, 2003; pp. 583–610.

29. Shi, M.; Wang, Z. *Principle and Design of Heat Exchangers*, 4th ed.; Southeast University Press: Nanjing, China, 2009; pp. 53–107.

30. Yue, S.; Wang, Y.; Wang, H. Design and optimization of tandem arranged cascade in a transonic compressor. *J. Therm. Sci.* **2018**, *27*, 349–358. [CrossRef]

31. Wen, J.; Yang, H.; Jian, G.; Tong, X.; Li, K.; Wang, S. Energy and cost optimization of shell and tube heat exchanger with helical baffles using Kriging metamodel based on MOGA. *Int. J. Heat Mass Transf.* **2016**, *98*, 29–39. [CrossRef]

32. Wen, J.; Yang, H.; Tong, X.; Li, K.; Wang, S.; Li, Y. Optimization investigation on configuration parameters of serrated fin in plate-fin heat exchanger using genetic algorithm. *Int. J. Therm. Sci.* **2016**, *101*, 116–125. [CrossRef]

33. Wen, J.; Yang, H.; Tong, X.; Li, K.; Wang, S.; Li, Y. Configuration parameters design and optimization for plate-fin heat exchangers with serrated fin by multi-objective genetic algorithm. *Energ. Convers. Manag.* **2016**, *117*, 482–489. [CrossRef]

34. Deb, K.; Pratap, A.; Agarwal, S.; Meyarivan, T. A fast and elitist multiobjective genetic algorithm: NSGA-II. *IEEE Trans. Evol. Comput.* **2002**, *6*, 182–197. [CrossRef]

35. Kays, W.M.; London, A.L. *Compact Heat Exchangers*, 3rd ed.; McGraw-Hill: New York, NY, USA, 1984; pp. 11–78.

36. Dong, Q.; Zhang, Y. *Petrochemical Equipment Design and Selection Manual: Heat Exchanger*; Chemistry Industry Publisher: Beijing, China, 2008; pp. 66–115.

 © 2019 by the authors. Licensee MDPI, Basel, Switzerland. This article is an open access article distributed under the terms and conditions of the Creative Commons Attribution (CC BY) license (http://creativecommons.org/licenses/by/4.0/).

Article

Improving Performance of Simplified Computational Fluid Dynamics Models via Symmetric Successive Overrelaxation

Vojtěch Turek

Sustainable Process Integration Laboratory–SPIL, NETME Centre, Faculty of Mechanical Engineering, Brno University of Technology–VUT Brno, Technická 2, 616 00 Brno, Czech Republic; turek@fme.vutbr.cz

Received: 23 May 2019; Accepted: 21 June 2019; Published: 25 June 2019

Abstract: The ability to model fluid flow and heat transfer in process equipment (e.g., shell-and-tube heat exchangers) is often critical. What is more, many different geometric variants may need to be evaluated during the design process. Although this can be done using detailed computational fluid dynamics (CFD) models, the time needed to evaluate a single variant can easily reach tens of hours on powerful computing hardware. Simplified CFD models providing solutions in much shorter time frames may, therefore, be employed instead. Still, even these models can prove to be too slow or not robust enough when used in optimization algorithms. Effort is thus devoted to further improving their performance by applying the symmetric successive overrelaxation (SSOR) preconditioning technique in which, in contrast to, e.g., incomplete lower–upper factorization (ILU), the respective preconditioning matrix can always be constructed. Because the efficacy of SSOR is influenced by the selection of forward and backward relaxation factors, whose direct calculation is prohibitively expensive, their combinations are experimentally investigated using several representative meshes. Performance is then compared in terms of the single-core computational time needed to reach a converged steady-state solution, and recommendations are made regarding relaxation factor combinations generally suitable for the discussed purpose. It is shown that SSOR can be used as a suitable fallback preconditioner for the fast-performing, but numerically sensitive, incomplete lower–upper factorization.

Keywords: computational fluid dynamics; symmetric successive overrelaxation; preconditioning; performance

---

1. Introduction

In engineering practice, it is often the case that process equipment is designed according to various rules of thumb. No optimization is generally done and, at best, a single computational fluid dynamics (CFD) simulation is carried out to verify that the design meets the key requirements of the future operator of the apparatus. This means that suboptimal designs or solutions, potentially leading to operating problems, are not uncommon.

One of the ways to remedy the situation is to use simplified CFD models. In spite of them not being as accurate as the standard CFD models, it has been shown [1] that they can provide useful quantitative information. What is more, these models feature significantly shorter computational times and their application in optimization algorithms is therefore much less cumbersome. To obtain solutions even faster, however, the numerical methods used to solve the underlying linear systems of equations can also be preconditioned. This means that instead of solving the original linear system

$$\mathbf{A}\mathbf{x} = \mathbf{b}, \tag{1}$$

where $\mathbf{A}$ is the coefficient matrix, $\mathbf{x}$ the solution vector, and $\mathbf{b}$ the right-hand side vector, one considers the system

$$\mathbf{M}^{-1}\mathbf{A}\mathbf{x} = \mathbf{M}^{-1}\mathbf{b}. \tag{2}$$

here, $\mathbf{M}$ denotes the preconditioning matrix such that $\mathbf{M}^{-1}\mathbf{A}$ has a smaller condition number than $\mathbf{A}$ and, therefore, linear system (2) features better convergence. It should also be noted that $\mathbf{M}^{-1}\mathbf{A}$ often is not formed explicitly but, instead, $\mathbf{M}\mathbf{u} = \mathbf{v}$ is solved for various auxiliary vectors $\mathbf{u}$ and $\mathbf{v}$ within the numerical solution method itself.

There are many different preconditioning techniques (i.e., ways to choose $\mathbf{M}$) available, and the selection of the best one for a particular purpose depends mainly on the type of equation that is being solved and the employed ordering of the variables. However, the most commonly used techniques are likely various flavors of incomplete lower–upper factorization (ILU) [2] (these were numerically investigated by Chapman et al. [3]) and symmetric successive overrelaxation (SSOR) [4]. Although SSOR was originally intended for symmetric matrices, it was shown [5] to also work when the matrices are not symmetric. Assuming the splitting $\mathbf{A} = \mathbf{D} + \mathbf{L} + \mathbf{U}$, where $\mathbf{D}$ is the diagonal of $\mathbf{A}$ and $\mathbf{L}$ and $\mathbf{U}$ its strictly lower and upper triangular parts, respectively, the SSOR preconditioning technique is applied within the numerical solution method as two SOR sweeps using different values of ω. The iteration process for auxiliary vectors $\mathbf{u}$, $\mathbf{v}$ (depend on the actual solution method used) can then be written as

$$\begin{aligned}
\mathbf{u}^{(k+1/2)} &= (\mathbf{D} + \omega_F\mathbf{L})^{-1}[(1-\omega_F)\mathbf{D} - \omega_F\mathbf{U}]\mathbf{u}^{(k)} + \omega_F(\mathbf{D} + \omega_F\mathbf{L})^{-1}\mathbf{v} \text{ (forward sweep)} \\
\mathbf{u}^{(k+1)} &= (\mathbf{D} + \omega_R\mathbf{U})^{-1}[(1-\omega_R)\mathbf{D} - \omega_R\mathbf{L}]\mathbf{u}^{(k+1/2)} + \omega_R(\mathbf{D} + \omega_R\mathbf{U})^{-1}\mathbf{v}, \text{ (backward sweep)}
\end{aligned} \tag{3}$$

where ω_F and ω_R denote the corresponding forward and backward relaxation factors. In other words, direct application of the inverse of the preconditioning matrix, $\mathbf{M}^{-1}$, is replaced by preconditioned fixed point iteration.

The advantages of SSOR are evidenced by the existence of a multitude of papers discussing improved versions of this technique or its extensions to various specific applications. Bai [6] studied SSOR-like preconditioners for non-Hermitian positive definite linear systems, for which the respective matrix was either Hermitian-dominant or skew–Hermitian-dominant. The paper also discussed the results of numerical implementations, showing that Krylov subspace iteration methods, when accelerated using SSOR-like preconditioners, are efficient solvers for classes of non-Hermitian positive definite linear systems. A "shifted" version of SSOR for non-Hermitian positive definite linear systems with a dominant Hermitian part was proposed by Tan [7]. Zhang [8], on the other hand, introduced an SSOR-like preconditioner for saddle point problems with a dominant skew-Hermitian part. A class of hybrid preconditioning methods for accelerated solution of saddle point problems was discussed by Wang [9], while Chen et al. [10] proposed a version of SSOR suitable for preconditioning of large dense complex linear systems arising from three-dimensional electromagnetic scattering. Wu and Li [11] introduced a modified SSOR technique for the solution of Helmholtz equations.

Preconditioning can also be done block-wise. This was discussed, e.g., by Zhang and Cheng [12] in terms of large sparse saddle point problems, and by Huang and Lu [13], who focused on SSOR block preconditioners applied in image restoration. Because, in fact, preconditioning means obtaining an easily invertible approximation of the original matrix, one can also use SSOR for just this purpose as shown, for example, by Meng et al. [14] in the context of fast recovery of density 3D data from gravity data. Similarly, a massively-parallel GPU implementation of the conjugate gradient method, which uses the approximate inverse matrix derived from SSOR as the preconditioning matrix, was proposed by Helfenstein and Koko [15].

Performance comparisons of SSOR and other, simpler preconditioning techniques were presented, e.g., by Meyer [16], who focused on genomic evaluation and by Sanjuan et al. [17], who used SSOR to accelerate parallel wind field calculations. In the latter paper, the authors also evaluated a new, reordered sparse matrix storage format and showed that this format can markedly shorten computational time. This confirms the earlier findings of Duff and Meurant [18], who investigated the effect of ordering on the convergence of the conjugate gradient method preconditioned, among others, using SSOR, or DeLong and Ortega [19], who focused on parallel implementations of SOR in terms of natural and multicolor orderings. Chen et al. [20], on the other hand, proposed a novel reordering technique for SSOR approximate inverse preconditioner, used together with a GPU-accelerated conjugate gradient solver, which should maximize the coalescing of global memory accesses.

Many SSOR-like, a priori preconditioned numerical solution methods using different splittings of the coefficient matrix have also been proposed for various types of problems. A three-parameter extension of the SSOR method intended for singular saddle point problems, which commonly arise, e.g., in fluid dynamics, was proposed by Li and Zhang [21]. A differently accelerated generalized three-parameter method for both singular and non-singular problems was introduced by Pan [22]. Similarly, a three-parameter unsymmetric SOR method for such saddle point problems was proposed, for example, by Liang and Zhang [23], who also discussed the choice of optimal values of the parameters. Many different SSOR-like methods are available for augmented systems, as well. Wang and Huang [24] introduced a four-parameter method, Louka and Missirlis [25] introduced a five-parameter extrapolated form of SSOR, and Najafi and Edalatpanah [26] introduced an improved version of the modified SSOR method for large sparse augmented systems, proposed earlier by Darvishi and Hessari [27]. Another improved SSOR method intended for the solution of augmented systems was proposed by Salkuyeh et al. [28]. In the case of complex systems, one can use, for instance, the accelerated method by Huang et al. [29], that is, an accelerated version of the method by Edalatpour et al. [30], in which the solution vector is split into two subvectors and different relaxation factors are used when solving for each of them. Likewise, one can employ the preconditioned variant of the generalized SSOR method by Hezari et al. [31] or the method by Salkuyeh et al. [32], which solves a real system obtained from the original, complex one. Block linear systems can be solved, e.g., using the block-preconditioned SSOR method by Pu and Wang [33].

The majority of the papers mentioned above discuss convergence (or at least semi-convergence) of the proposed methods, and many also include some information on the optimal selection of the relaxation factors. Kushida [34] focused on the estimation of convergence of the original SSOR preconditioner via a condition number, while general discussion related to SSOR-like methods for non-Hermitian positive definite linear systems was published in [35]. Augmented systems were addressed, for example, by Wang and Huang [36]. Similarly, there are papers focusing on convergence and optimal selection of the relaxation factors in the case of methods for block 2×2 linear systems [37], saddle point problems [38], parallel SSOR implementations [39], the Poisson equation [40], etc. In all these cases, however, convergence was investigated via spectral analysis, which is often prohibitively expensive [41]. The present paper, focusing on fast estimation of suitable SSOR relaxation factors in engineering practice, therefore, investigates the convergence experimentally using several different simplified 3D CFD flow models. The suitability of specific combinations of relaxation factors is assessed on the basis of mean computational times needed to reach converged steady-state solutions. The best-performing combinations of relaxation factors are then given together with the obtained relaxation factor trends.

2. Materials and Methods

Three different flow systems, with both the "U" and the "Z" flow arrangements ("U": outlet on the same side of the flow system as the inlet, "Z": outlet on the opposite side), were used to generate test cases. Moreover, in two of these three flow systems, the mesh fineness was also varied (coarser and finer mesh). This yielded ten flow system configurations in total, with simplified, cuboid cell-only meshes of different sizes ranging from ~6000 cells to ~41,000 cells (see Table 1). The meshes were generated automatically by the employed benchmarking software (see further) using the key parameters listed in Table 1. Due to the cuboid nature of the meshes, cell sizes were, in all computational domains, governed primarily by how many cell faces comprised a tube cross section (coarser mesh: 1 face only, finer mesh: 2×2 faces) and by the utilized cell growth factor. Sample meshes are shown in Figure 1.

Table 1. Flow system parameters: W, L, and H denote width, length, and height, respectively, p_r row pitch, and p_t tube pitch.

Parameter	Flow System 1	Flow System 2	Flow System 3
Headers ($W \times L \times H$)	$40 \times 320 \times 40$ mm	$60 \times 340 \times 60$ mm	$100 \times 340 \times 100$ mm
Inlet/outlet region (L)	60 mm	60 mm	60 mm
Tube bundle	2 inline rows, 20 tubes/row, $p_r = p_t = 15.6$ mm	3 staggered rows (45°), 10 tubes/row, $p_r = 15.6$ mm, $p_t = 2p_r$	5 staggered rows (45°), 10 tubes/row, $p_r = 15.6$ mm, $p_t = 2p_r$
Tubes	$\varnothing 10$ mm $\times 500$ mm	$\varnothing 10$ mm $\times 500$ mm	$\varnothing 10$ mm $\times 500$ mm
Flow arrangement	"U" or "Z"	"U" or "Z"	"U" or "Z"
Mesh	coarser: ~6000 cells, finer: ~25,000 cells	coarser: ~8000 cells, finer: ~41,000 cells	coarser: ~16,000 cells

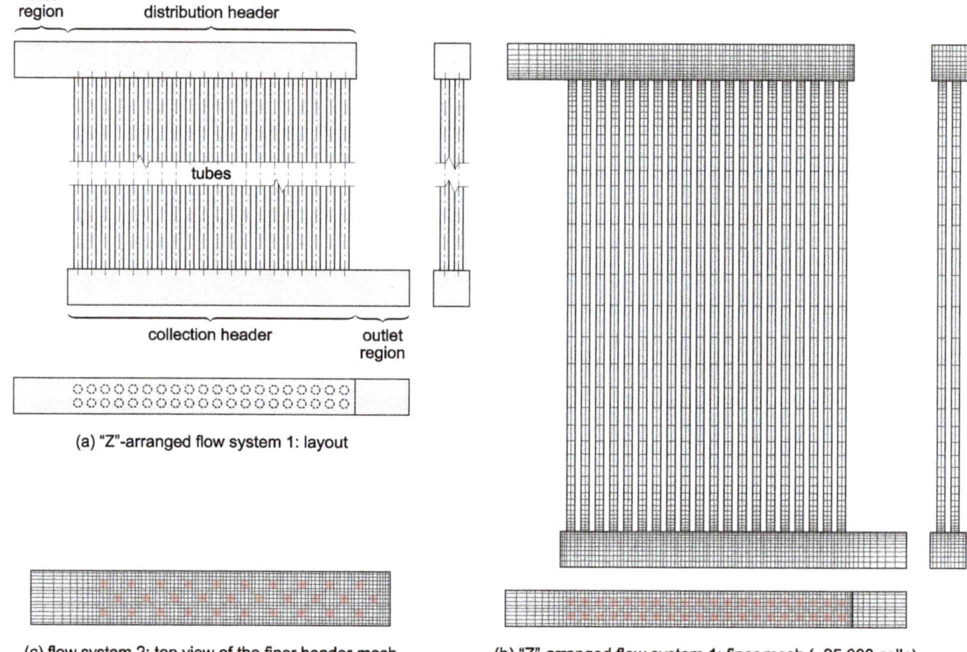

Figure 1. Sample simplified (cuboid cell-only) meshes generated automatically by the benchmarking software using the growth factor of 1.15; clockwise from top left: (**a**) layout of the "Z"-arranged flow system 1 from Table 1; (**b**) the corresponding finer mesh (~25,000 cells, groups of red cells in the top view denote locations of tube cross-sections); (**c**) top view of the finer mesh of the distribution header from flow system 2 (inlet region being on the left); other parts of this mesh, as well as the remaining meshes, were generated analogously.

As for boundary conditions, 0.5 kg/s of water at 300 K was fed into the inlet while pressure at the outlet was set to 101,325 Pa. All walls were adiabatic except for tube walls, where a specific heat flux of 15 kW/m^2 was set when the energy equation was enabled. Steady state simulations were carried out using the same CFD setup as in [42], that is, the SIMPLEC pressure-velocity coupling [43] was employed together with the Power Law discretization scheme [44]. Standard scaled residual limits, i.e., 10^{-3} for continuity and momentum and 10^{-6} for energy, were used. Only the natural ordering of the variables was considered in this study.

The SSOR preconditioning technique was paired with two widely used numerical solution methods, which were shown by an earlier study [42] to perform very well in simplified 3D CFD models. The conjugate gradient (CG) numerical solution method [45] was employed in the pressure correction equation. The momentum and energy equations, on the other hand, were solved using the bi-conjugate gradient stabilized numerical method with the minimization of residuals over L-dimensional subspaces (BiCGstab(L)) [46], with $L = 1, 2,$ or 3. Performance of the ILU preconditioner (which is efficient, but the construction of the respective preconditioning matrix may not always be possible) was taken as the baseline.

The SSOR-preconditioned numerical solution methods were tested with various tuples of ω_F, $\omega_R = 0.1, 0.2, \dots, 1.8, 1.9$ in successive steps, depending on the obtained results. Promising combinations of ω_F and ω_R were then taken as pivots, and their square neighborhoods were tested further—that is, all combinations of ω_F, ω_R with the respective values being $\omega - 0.04$, $\omega - 0.02$, ω, $\omega + 0.02$, and $\omega + 0.04$ were evaluated except for the original pivot point (ω_F, ω_R). In order to be able to compare SSOR to the baseline (ILU), all the ten meshes were also evaluated using the ILU-preconditioned combinations of solution methods. In total, 50,620 CFD model setups were tested.

The same benchmarking simplified 3D CFD Java software application was used as in [42], and, therefore, the reader is kindly referred to this paper for details (please note that the software is not publicly available). The benchmarking procedure itself was almost the same, as well, with the only difference being that the numbers of warm-up and test runs were smaller for the larger meshes (see Table 2). Such a measure was necessary to keep the times required to complete the respective benchmarks within reasonable bounds. This did not introduce any problems, because with larger cell counts all Java initializations and compilations had been finished within much less warm-up runs, and, therefore, it was not needed to carry out many of them before the timing phase. Mean test-run computational time was then taken as the final performance metric. Unlike in [42], however, only one machine (Intel Xeon E5 2698 v4 CPU, 128 GB RAM) was used instead of two largely disparate ones. The reason for this simplification was that, as shown in the respective paper, single-core computational times proved to be virtually identical, irrespective of whether the machine was a high-performance server or a regular laptop with an ultra-low voltage CPU. Please see the paragraph titled Supplementary Materials on how to obtain the data set containing all the mean computational times together with other relevant information.

Table 2. Numbers of warm-up and test runs and test case limits.

Mesh Size (Cells)	Warm-Up Runs	Test Runs	Iteration Limit	Computational Time Limit
~6000	30	50	1000	1800 s
~8000	30	50	1000	1800 s
~16,000	20	40	1500	2700 s
~25,000	10	30	2000	3600 s
~41,000	10	30	2000	3600 s

Because this study targeted fast computation, two kinds of limits were set in the solution process as detailed in Table 2. The first one concerned the number of CFD solver iterations, while the second one applied to the actual computational time. Any combination of numerical solution methods and preconditioning techniques which exceeded at least one of these two limits was marked as failing to reach a solution. Additionally, since robustness is one of the factors that must be considered when evaluating the suitability of numerical solution methods, no user interventions (e.g., changes to the internal residual limits of the numerical methods, CFD relaxation factors, etc.) were allowed during a solution process.

3. Results

This section is split into four parts. The first part presents the results pertaining to flow-only simulations. The second part discusses flow and energy transport simulations, in which only the energy equation was solved using SSOR-preconditioned numerical methods. The third part, on the other hand, summarizes data obtained via benchmarks, where SSOR was used for both the momentum and the energy equations. The last part then compares the SSOR-related data with results corresponding to the cases where solely the ILU preconditioning technique was used.

3.1. Flow-Only Simulations with SSOR-Preconditioned Momentum Equations

As mentioned above, the pressure correction equation was always solved using CG. This solution method was first coupled with SSOR, while the momentum equations were solved using BiCGstab(3):ILU. Within the several initial benchmarks, however, it became clear that CG:SSOR is wholly unsuitable. No matter the actual relaxation factors ω_F and ω_R, the majority of test cases either failed (mostly due to divergence) or resulted in very long computational times. Those setups which did not fail featured $\omega_F \approx \omega_R$ and, what is more, their amount decreased with an increasing mesh size, as shown in Figure 2. This was most probably a consequence of the fact that SSOR is sensitive to the ordering of variables. Only the ILU preconditioning technique was, therefore, used for the pressure correction equation, from then on.

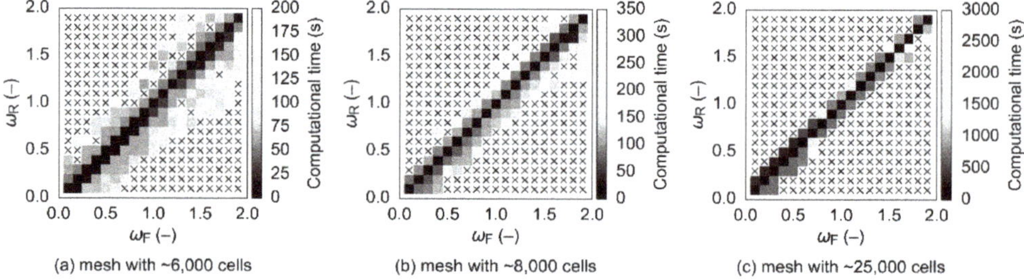

Figure 2. Mean computational times for three different meshes and various combinations of ω_F and ω_R, the pressure correction equation was solved using CG:SSOR and the momentum equations using BiCGstab(3):ILU; diagonal crosses indicate failing combinations of ω_F and ω_R. Please note that the colorbars have been adjusted so that the best (i.e., the most relevant) combinations of ω_F, ω_R are easily identifiable. Further, due to the necessary colorbar ranges, the respective lower limits have been set to zero even though the data start at higher values.

The shortest mean computational times obtained with BiCGstab(2) and BiCGstab(1) instead of BiCGstab(3) were, on average, ~80% and ~260% longer, respectively. Conversely, stability of the solution process was greater with these two methods, and fewer combinations of ω_F and ω_R resulted in failures (again, mostly because of divergence; see Figure 3). It is also evident from the figure that the best-performing combinations were around $\omega_F = \omega_R = 1.0$, while with ω_F below 0.5 or above 1.5 the computational times were much longer, or solution failures occurred. As for the CFD setup involving BiCGstab(3) in particular, mean computational times started at 2.56 s for the smallest mesh and 117.33 s when the largest mesh was used.

The solution behavior mentioned above was also observed for the larger meshes. Only BiCGstab(3):SSOR was, therefore, used to solve the momentum equations in the following flow-only benchmarks because of its superior performance. To generate these, the combinations of ω_F and ω_R obtained for the ten flow systems were sorted by mean computational time and the respective ordered sets were then used to find the most common combinations yielding the most favorable computational times. In other words, the best tuples (ω_F, ω_R) were taken as pivot points whose square neighborhoods were evaluated further.

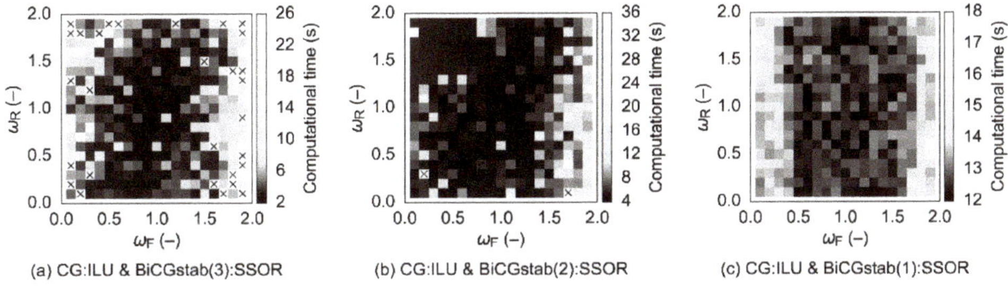

(a) CG:ILU & BiCGstab(3):SSOR   (b) CG:ILU & BiCGstab(2):SSOR   (c) CG:ILU & BiCGstab(1):SSOR

Figure 3. Mean computational times for the smallest mesh and the cases when the pressure correction equation was solved using CG:ILU and the momentum equations using BiCGstab(3):SSOR, BiCGstab(2):SSOR, and BiCGstab(1):SSOR; diagonal crosses indicate failing combinations of ω_F and ω_R. Please note that the colorbars have been adjusted so that the best (i.e., the most relevant) combinations of ω_F, ω_R are easily identifiable.

Example plots resulting from such evaluations are shown in Figure 4. Both pertain to the same mesh—the smallest one in this particular case. The plot on the left (Figure 4a) shows mean computational times for all the pivots and their respective neighborhoods. The plot on the right (Figure 4b) displays a cropped area corresponding to ω_F, $\omega_R \in [0.5, 1.5]$, where the best-performing combinations of relaxation factors are located. The dotted lines in both these plots represent the trend obtained using the standard weighted least squares method. Because the goal was to minimize computational time, the weights for individual combinations (ω_F, ω_R) were calculated as $w_i = (t_{min}/t_i)^4$, where t_{min} denotes the minimum computational time observed with a specific mesh and t_i denotes the mean computational time corresponding to the respective (i-th) combination of relaxation factors evaluated using this mesh. The fourth power of the computational time ratio instead of just the ratio itself was used to adequately limit the influence of combinations that yielded solutions in longer time frames. Weights for combinations leading to solution failures were set to zero.

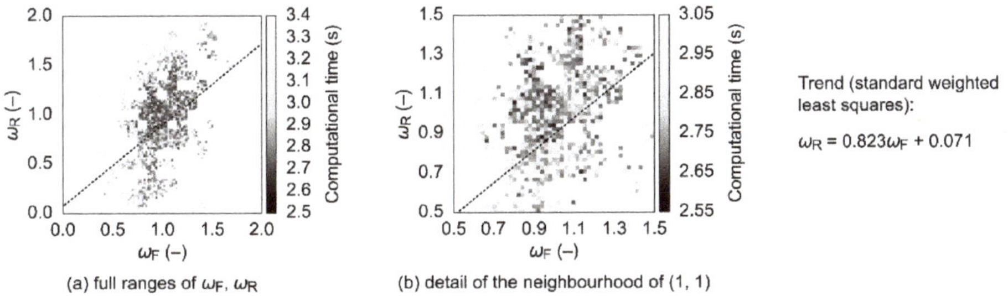

(a) full ranges of ω_F, ω_R (b) detail of the neighbourhood of (1, 1)

Figure 4. Mean computational times for the neighborhoods of pivot points corresponding to the smallest mesh; the pressure correction equation was solved using CG:ILU and the momentum equations using BiCGstab(3):SSOR; please note that, for the sake of clarity, the time ranges have been severely limited in both plots. Again, the colorbars have been adjusted so that the best (i.e., the most relevant) combinations of ω_F, ω_R are easily identifiable.

Trends for all the mesh sizes as well as the overall relaxation factor trends are listed in Table 3 and shown in Figure 5. Although the R^2 values are relatively low, this is caused by the fact that the data featured many less-relevant points scattered over the $(0, 2) \times (0, 2)$ relaxation factor domain, which were assigned small, but still non-zero weights. In any case, the standard errors for the trend coefficients are quite reasonable, and it is obvious that all the trends are very similar. Considering the actual values of the coefficients a and b and the fact that the best combinations of relaxation factors featured $\omega_F \approx 0.9$, it follows that, when solving the momentum equations in flow-only scenarios, both SSOR sweeps should be slightly underrelaxed to gain the shortest computational time.

Table 3. Relaxation factor trends, $\omega_R = a\omega_F + b$, for individual mesh sizes and the overall relaxation factor trend for the solution of momentum equations using BiCGstab(3):SSOR; R^2 denotes the coefficient of determination and $SE(a)$ and $SE(b)$ the standard error values for the coefficients a, b.

Mesh Size (Cells)	a	b	R^2	$SE(a)$	$SE(b)$
~6000	0.823	0.071	0.479	0.014	0.009
~8000	0.843	0.094	0.457	0.015	0.007
~16,000	0.888	0.011	0.795	0.013	0.003
~25,000	0.798	0.051	0.593	0.019	0.007
~41,000	0.898	0.028	0.754	0.015	0.007
Overall trend	0.861	0.056	0.584	0.007	0.003

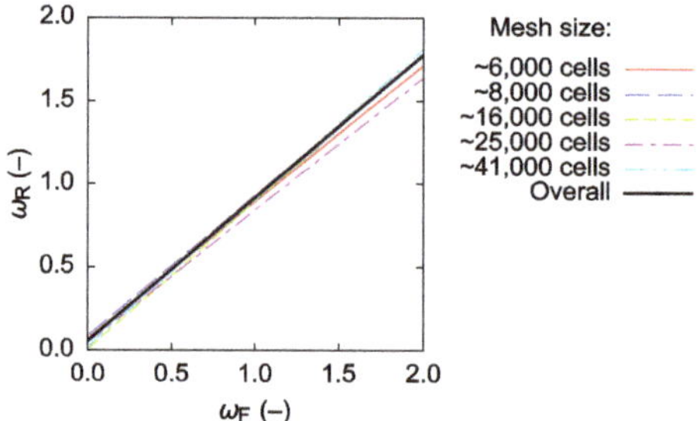

Figure 5. Relaxation factor trends for the solution of momentum equations using BiCGstab(3):SSOR; the overall trend was obtained using the merged data set and identical weights.

3.2. Flow & Energy Transport Simulations with SSOR-Preconditioned Energy Equation

Based on the solution behavior observed in the flow-only scenarios, only CG:ILU was used to solve the pressure correction equation in the flow and energy transport simulations. Momentum equations were also preconditioned only with ILU, while SSOR was used just for the energy equation. Various combinations of BiCGstab(L) for the momentum and energy equations were tested first, and the two most suitable combinations were then evaluated in detail using square neighborhoods of promising relaxation factor tuples (i.e., the pivot points).

The best results overall were obtained using BiCGstab(3) and BiCGstab(1) for the momentum equations and the energy equation, respectively (see Figure 6). This setup was relatively robust, most probably because of the better stability and smoothness of convergence resulting from the use of BiCGstab(1). Figure 7 shows the second-best combination, featuring only BiCGstab(2). On average, this setup was ~38% slower, and more combinations of ω_F and ω_R resulted in solution failures. It can also be seen that all the suitable relaxation factor tuples were in a relatively small neighborhood of (1, 1). Furthermore, with the SSOR-preconditioned energy equation, a significantly larger percentage of failures than before was due to slow convergence (that is, the respective iteration limits were exceeded).

The data sets mentioned above were then combined with data sets obtained by evaluating square neighborhoods of the promising tuples of ω_F and ω_R to get the corresponding relaxation factor trends. Again, the standard weighted least squares method was used with the weights being calculated in the same manner as before. Because the best setup and the second-best one (featured in Figures 6 and 7) were, at least in some cases, on par, the trends were calculated for both of them (see Table 4 and Figure 8). It is of note here that all benchmarks involving the largest mesh and BiCGstab(1) had failed. This suggests that the respective solution method simply is too slow when combined with SSOR. In any case, the best results were obtained with ω_F around 1.2 or 1.1 when BiCGstab(1) or BiCGstab(2) were used, respectively. This means that, given the calculated relaxation factor trends, ω_R should also be a little above 1.0 (i.e., it is best to slightly overrelax both SSOR sweeps when solving the energy equation).

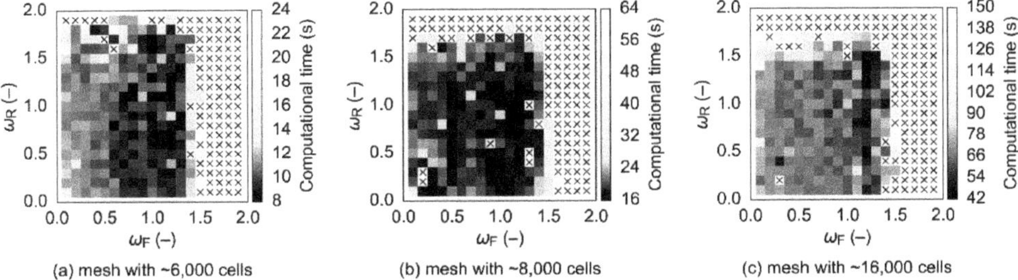

(a) mesh with ~6,000 cells   (b) mesh with ~8,000 cells   (c) mesh with ~16,000 cells

Figure 6. Mean computational times for three different meshes and various combinations of ω_F and ω_R, CG:ILU was used to solve the pressure correction equation, BiCGstab(3):ILU was used for the momentum equations, and BiCGstab(1):SSOR was used for the energy equation; diagonal crosses indicate failing combinations of ω_F and ω_R. Please note that the colorbars have been adjusted so that the best (i.e., the most relevant) combinations of ω_F, ω_R are easily identifiable.

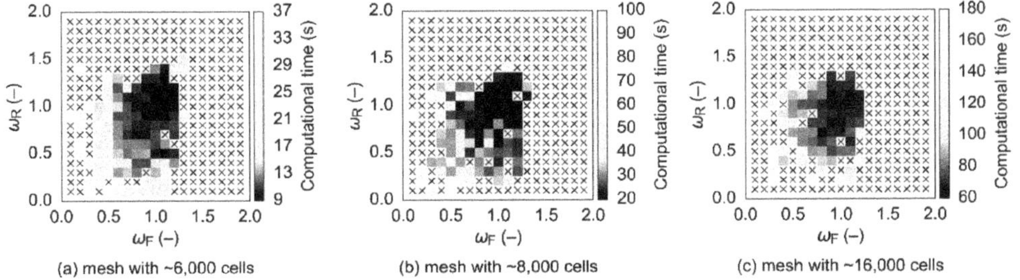

(a) mesh with ~6,000 cells   (b) mesh with ~8,000 cells   (c) mesh with ~16,000 cells

Figure 7. Mean computational times corresponding to the meshes from Figure 6 and the cases when the momentum equations were solved using BiCGstab(2):ILU, and the energy equation was solved using BiCGstab(2):SSOR; diagonal crosses indicate failing combinations of ω_F and ω_R. Please note that the colorbars have been adjusted so that the best (i.e., the most relevant) combinations of ω_F, ω_R are easily identifiable.

Table 4. Relaxation factor trends, $\omega_R = a\omega_F + b$, for individual mesh sizes and the overall relaxation factor trends for the two discussed setups using either BiCGstab(1):SSOR or BiCGstab(2):SSOR to solve the energy equation; R^2 denotes the coefficient of determination, $SE(a)$ and $SE(b)$ denote the standard error values for the coefficients a, b, and "n/a" denotes the fact that all the respective benchmarks had failed, and thus the trend could not be obtained.

Mesh Size (Cells)	BiCGstab(1)					BiCGstab(2)				
	a	b	R^2	$SE(a)$	$SE(b)$	a	b	R^2	$SE(a)$	$SE(b)$
~6000	0.867	0.047	0.555	0.027	0.009	0.960	0.007	0.929	0.008	0.004
~8000	0.762	0.120	0.564	0.023	0.013	0.971	0.005	0.936	0.007	0.003
~16,000	0.717	0.069	0.512	0.037	0.011	0.995	0.003	0.944	0.006	0.002
~25,000	0.905	0.012	0.888	0.009	0.002	0.964	0.012	0.922	0.013	0.006
~41,000	n/a	n/a	n/a	n/a	n/a	0.994	0.001	0.940	0.013	0.001
Overall trend	0.866	0.042	0.676	0.009	0.003	0.972	0.005	0.939	0.004	0.001

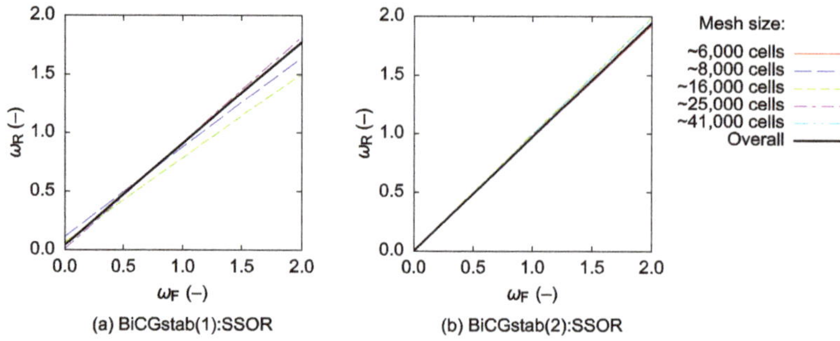

Figure 8. Relaxation factor trends for the two discussed setups using either BiCGstab(1):SSOR or BiCGstab(2):SSOR to solve the energy equation; the overall trends were obtained using the merged data sets and identical weights.

3.3. Flow & Energy Transport Simulations with SSOR-Preconditioned Momentum and Energy Equations

The last set of SSOR benchmarks involved both the momentum and the energy equations being preconditioned using this technique. However, because of a significantly larger relaxation factor domain (four factors had to be chosen instead of two), only ω_F, $\omega_R = 0.7, 0.8, \ldots , 1.3$ were evaluated, i.e., only the subdomain where the best-performing relaxation factor quadruples were expected to lie. Additionally, only the "Z"-arranged flow system meshes, and the two setups identified in Section 3.2 as the most promising ones, were considered. Such a reduction led to a decrease in the number of combinations to be evaluated from more than 2.6 million (10 meshes × 2 setups × 130,321 factor quadruples) to ~24 thousand (5 meshes × 2 setups × 2,401 factor quadruples). This still provided enough information to get a general sense of how solution processes would likely behave.

The respective benchmarks generally resulted in computational times and solution failure percentages comparable to those reached when just the energy equation was preconditioned using SSOR. As before, the failures mostly occurred due to slow convergence or—less often—because of divergence. Only with rare combinations of SSOR relaxation factors were the solution processes so slow that the respective time limit was exceeded.

The best-performing combinations of relaxation factors are listed in Table 5. It can be seen that with almost all meshes, the setup involving only BiCGstab(2) resulted in markedly longer computational times. Additionally, it should be noted that there were other combinations of factors providing similar numerical performance, but all of them were clustered around the values mentioned in the table.

Table 5. Combinations of relaxation factors resulting in the shortest mean computational times when both the momentum and the energy equations were preconditioned using SSOR; setup "B3/B1" denotes the case when BiCGstab(3) was used to solve the momentum equations and BiCGstab(1) the energy equation, while setup "B2/B2" corresponds to only BiCGstab(2) being used for both these equation types.

Mesh Size (Cells)	Setup	Momentum		Energy		Mean Computational Time
		ω_F	ω_R	ω_F	ω_R	
~6000	B3/B1	1.3	1.1	1.2	0.7	6.87 s
	B2/B2	1.0	0.9	0.9	1.2	8.34 s
~8000	B3/B1	1.0	1.0	0.8	1.0	14.46 s
	B2/B2	1.3	1.3	1.0	1.1	22.58 s
~16,000	B3/B1	1.3	1.3	1.0	0.7	81.65 s
	B2/B2	0.8	0.7	1.0	1.3	57.96 s
~25,000	B3/B1	1.1	0.7	1.2	1.2	99.16 s
	B2/B2	1.3	1.3	1.0	1.2	134.85 s
~41,000	B3/B1	1.1	0.8	1.2	0.8	325.18 s
	B2/B2	1.0	0.8	1.0	1.1	542.41 s

Visualizing the obtained data in one plot per data set is not possible because that would require four-dimensional plots. One could, however, fix the momentum relaxation factor tuple to, e.g., the values from the best-performing quadruple mentioned in Table 5, and then plot the corresponding two-dimensional energy relaxation factor map (or vice versa). Examples of such plots are shown in Figures 9 and 10.

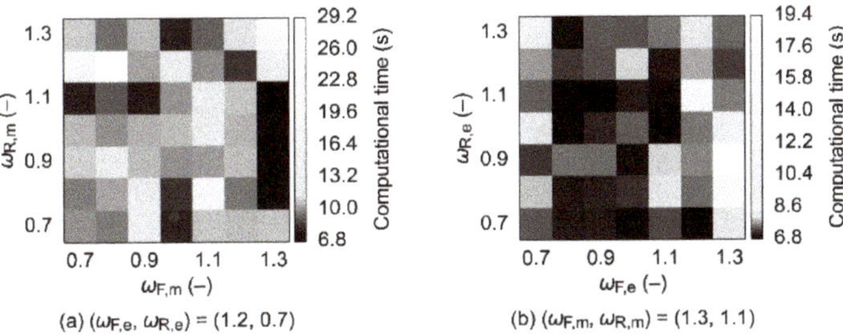

(a) $(\omega_{F,e}, \omega_{R,e}) = (1.2, 0.7)$ (b) $(\omega_{F,m}, \omega_{R,m}) = (1.3, 1.1)$

Figure 9. Two-dimensional plots of mean computational times obtained for the smallest "Z"-arranged mesh with (**a**) the energy relaxation factor tuple fixed to $(\omega_F, \omega_R) = (1.2, 0.7)$ and (**b**) the momentum relaxation factor tuple fixed to $(\omega_F, \omega_R) = (1.3, 1.1)$; BiCGstab(3):SSOR was used to solve the momentum equations and BiCGstab(1):SSOR the energy equation. Please note that the colorbars have been adjusted so that the best (i.e., the most relevant) combinations of ω_F, ω_R are easily identifiable.

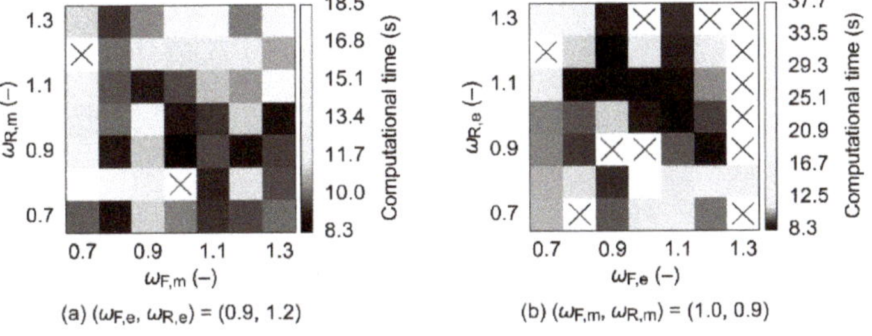

(a) $(\omega_{F,e}, \omega_{R,e}) = (0.9, 1.2)$ (b) $(\omega_{F,m}, \omega_{R,m}) = (1.0, 0.9)$

Figure 10. Two-dimensional plots of mean computational times obtained for the smallest "Z"-arranged mesh with (**a**) the energy relaxation factor tuple fixed to $(\omega_F, \omega_R) = (0.9, 1.2)$ and (**b**) the momentum relaxation factor tuple fixed to $(\omega_F, \omega_R) = (1.0, 0.9)$; BiCGstab(2):SSOR was used to solve both the momentum equations and the energy equation; diagonal crosses indicate failing combinations of ω_F and ω_R. Please note that the colorbars have been adjusted so that the best (i.e., the most relevant) combinations of ω_F, ω_R are easily identifiable.

Because, here, one must choose two relatively independent relaxation factor tuples, it is best to generate two trends for each combination of numerical solution methods. This can be done by "flattening" the four-dimensional data to two dimensions (while still considering all the data points). In other words, if one sought, e.g., the momentum relaxation factor trend, one would disregard the energy-related part of the relaxation factor quadruple and thus have multiple data points with different weights (calculated just as before) for each momentum relaxation factor tuple. The respective trends would then, again, be calculated via the standard weighted least squares method (see Tables 6 and 7 and Figures 11 and 12). From the results, it follows that for the momentum equations, the SSOR forward sweeps should generally be carried out with ω_F between ca. 1.1 and 1.3, while the backward sweeps should use $\omega_R \leq \omega_F$. As for the energy equation and BiCGstab(1), the forward sweep should, again, be slightly overrelaxed (ω_F up to ca. 1.2) and $\omega_R \leq \omega_F$, while with BiCGstab(2) ω_F should be around 1.0 (i.e., without any forward sweep relaxation) and $\omega_R \geq \omega_F$.

Table 6. Relaxation factor trends, $\omega_R = a\omega_F + b$, for individual mesh sizes and the overall relaxation factor trends for the setup where BiCGstab(3):SSOR was used to solve the momentum equations and BiCGstab(1):SSOR the energy equation; R^2 denotes the coefficient of determination and $SE(a)$ and $SE(b)$ the standard error values for the coefficients a, b.

Mesh Size (Cells)	Momentum					Energy				
	a	*b*	R^2	*SE(a)*	*SE(b)*	*a*	*b*	R^2	*SE(a)*	*SE(b)*
~6000	0.684	0.141	0.636	0.011	0.006	0.798	0.097	0.576	0.014	0.007
~8000	0.876	0.062	0.610	0.014	0.007	0.774	0.101	0.585	0.013	0.006
~16,000	0.957	0.000	0.995	0.001	0.000	0.985	0.000	0.937	0.005	0.000
~25,000	1.034	0.000	0.957	0.004	0.001	0.960	0.002	0.922	0.006	0.001
~41,000	0.931	0.001	0.914	0.006	0.001	0.892	0.001	0.905	0.006	0.001
Overall trend	0.936	0.013	0.893	0.003	0.001	0.949	0.011	0.884	0.003	0.001

Table 7. Relaxation factor trends, $\omega_R = a\omega_F + b$, for individual mesh sizes and the overall relaxation factor trends for the setup where BiCGstab(2):SSOR was used to solve both the momentum equations and the energy equation; R^2 denotes the coefficient of determination and $SE(a)$ and $SE(b)$ the standard error values for the coefficients a, b.

Mesh Size (Cells)	Momentum					Energy				
	a	*b*	R^2	*SE(a)*	*SE(b)*	*a*	*b*	R^2	*SE(a)*	*SE(b)*
~6000	0.864	0.054	0.786	0.009	0.005	0.963	0.031	0.856	0.008	0.004
~8,000	0.957	0.026	0.811	0.009	0.005	0.946	0.030	0.871	0.007	0.004
~16,000	1.007	0.007	0.845	0.009	0.004	0.980	0.016	0.904	0.007	0.003
~25,000	0.965	0.000	0.962	0.004	0.000	1.155	0.000	0.974	0.004	0.000
~41,000	0.993	0.002	0.930	0.006	0.001	1.007	0.001	0.961	0.004	0.001
Overall trend	0.965	0.011	0.882	0.003	0.001	0.990	0.010	0.923	0.003	0.001

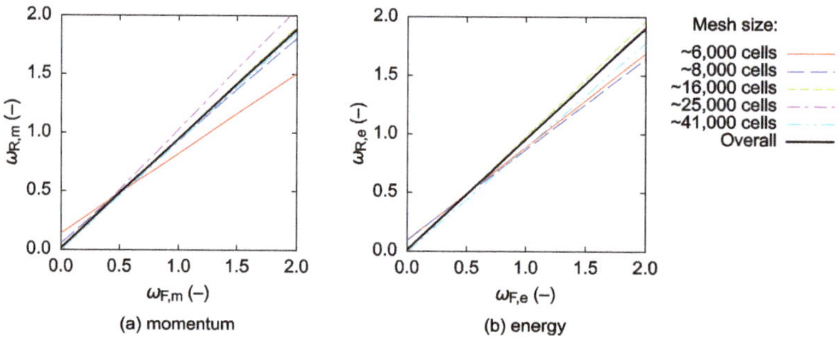

Figure 11. Relaxation factor trends for (**a**) momentum and (**b**) energy obtained with BiCGstab(3):SSOR and BiCGstab(1):SSOR as the solution methods for the momentum equations and the energy equation, respectively; the overall trends were obtained using the merged data sets and identical weights.

3.4. Comparison to ILU-Only Simulations

In order to be able to assess the potential benefit of using SSOR, the meshes listed in Table 1 were also evaluated via the two combinations of numerical solution methods from Table 5, but only with the ILU preconditioning technique being employed. The results, summarized in Table 8 and visually compared in Figure 13, suggest that utilizing SSOR leads to at least a ~23% increase (on average) in computational time.

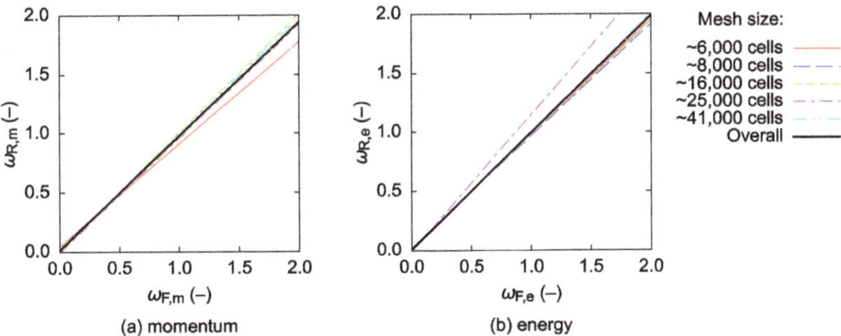

Figure 12. Relaxation factor trends for (**a**) momentum and (**b**) energy obtained with BiCGstab(2):SSOR as the solution method for both the momentum equations and the energy equation; the overall trends were obtained using the merged data sets and identical weights.

Table 8. Comparison of the mean computational times (s) obtained when solely the ILU preconditioning technique was used, and the overall best-case mean computational times reached with the momentum equations and/or the energy equation being preconditioned with SSOR.

Mesh Size (Cells)	ILU		SSOR		
	Momentum	Momentum & Energy	Momentum	Energy	Momentum & Energy
~6000	3.09	7.61	2.56	8.28	6.87
~8000	4.02	9.60	4.59	14.95	14.46
~16,000	12.93	30.92	15.60	42.13	57.96
~25,000	28.10	71.86	41.45	93.52	99.16
~41,000	119.33	294.51	117.33	368.33	325.18

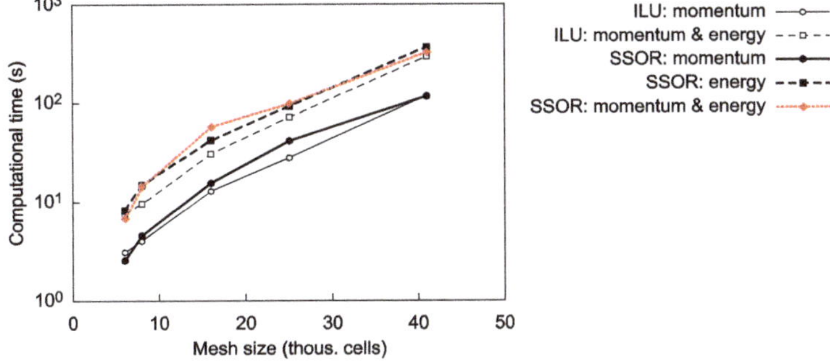

Figure 13. Comparison of the mean computational times obtained when solely the ILU preconditioning technique was used, and the overall best-case mean computational times reached with the momentum equations and/or the energy equation being preconditioned with SSOR.

4. Discussion

The aim of this study was to establish whether, in the case of simplified CFD models, the SSOR preconditioning technique can be a viable replacement for ILU. From the obtained data, it follows that SSOR should not be used in conjunction with CG to solve the pressure correction equation. When applied to the momentum and/or energy equations, computational times tend to be significantly longer even when the relaxation factors are chosen favorably (for the cases evaluated in this study, the increase was at least ~23% on average). However, because the SSOR preconditioning matrix can always be constructed, the respective techniques could be used in conjunction with ILU as a fallback option. From an engineering point of view, this would mean that ILU would be employed by default, and only in case of numerical issues would the CFD solver try to reach a converged solution using

SSOR. Such an approach would capitalize on the efficiency of ILU while maintaining reasonable numerical robustness due to the possibility of falling back to a technique with guaranteed existence of the preconditioning matrix. The resulting models would, ultimately, be much more suitable for implementation in optimization algorithms or for other use cases where large batches of simulations must be carried out without user intervention.

The best-performing combinations of numerical solution methods and SSOR forward and backward relaxation factors differ according to whether energy transport is included in the model or not. In the flow-only scenario, the momentum equations should preferably be solved using BiCGstab(3), with $\omega_F \approx 0.9$ and ω_R slightly less than ω_F, that is, both SSOR sweeps should be a little underrelaxed. Computational times obtained using other variants of BiCGstab(L) proved to be at least 80% longer. If also the energy transport is included and only the energy equation is preconditioned using SSOR, it is best to solve the momentum equations using BiCGstab(3) and the energy equation using BiCGstab(1). Here, both SSOR sweeps should be slightly overrelaxed, with $\omega_F \approx 1.2$ and $\omega_R \approx 1.1$. Similar performance can in some cases be obtained by employing BiCGstab(2) for both types of equations with the energy SSOR sweeps being overrelaxed using $\omega_F \approx \omega_R \approx 1.1$; however, a much greater solution failure probability can then be expected. If SSOR is utilized for both the momentum and the energy equations, then it is, again, preferable to use the combination of BiCGstab(3) and BiCGstab(1). The respective forward sweeps should be a little overrelaxed (ω_F between ca. 1.1 and 1.3 for the momentum, and up to ca. 1.2 for the energy), while the backward sweeps should feature ω_R slightly lower than ω_F. The best-case computational times obtained with BiCGstab(2) proved to be up to ~67% longer and, therefore, the use of this numerical solution method is discouraged in this scenario.

Supplementary Materials: The mean computational times together with other relevant information are available online at http://www.mdpi.com/1996-1073/12/12/2438/s1.

Funding: This research was funded by the Czech Republic Operational Programme Research, Development, and Education, Priority 1: Strengthening capacity for quality research, grant No. CZ.02.1.01/0.0/0.0/15_003/ 0000456 "Sustainable Process Integration Laboratory—SPIL".

Conflicts of Interest: The author declares no conflict of interest.

References

1. Turek, V.; Fialová, F.; Jegla, Z. Efficient flow modelling in equipment containing porous elements. *Chem. Eng. Trans.* **2016**, *52*, 487–492. [CrossRef]
2. Meijerink, J.A.; van der Vorst, A.H. An iterative solution method for linear systems of which the coefficient matrix is a symmetric M-matrix. *Math. Comput.* **1977**, *31*, 148–162. [CrossRef]
3. Chapman, A.; Saad, Y.; Wigton, L. High-order ILU preconditioners for CFD problems. *Int. J. Numer. Methods Fluids* **2000**, *33*, 767–788. [CrossRef]
4. Young, D.M. On the accelerated SSOR method for solving large linear systems. *Adv. Math.* **1977**, *23*, 215–271. [CrossRef]
5. Birken, P.; Gassner, G.; Haas, M.; Munz, C.-D. Preconditioning for modal discontinuous Galerkin methods for unsteady 3D Navier–Stokes equations. *J. Comput. Phys.* **2013**, *240*, 20–35. [CrossRef]
6. Bai, Z.-Z. On SSOR-like preconditioners for non-Hermitian positive definite matrices. *Numer. Linear Algebra* **2016**, *23*, 37–60. [CrossRef]
7. Tan, X.-Y. Shifted SSOR-like preconditioner for non-Hermitian positive definite matrices. *Numer. Algorithms* **2017**, *75*, 245–260. [CrossRef]
8. Zhang, J.-L. On SSOR-like preconditioner for saddle point problems with dominant skew-Hermitian part. *Int. J. Comput. Math.* **2019**, *96*, 782–796. [CrossRef]
9. Wang, Z.-Q. On hybrid preconditioning methods for large sparse saddle-point problems. *Linear Algebra Appl.* **2011**, *434*, 2353–2366. [CrossRef]
10. Chen, J.; Liu, Z.; Cao, N.; Yong, B. Shifted SSOR preconditioning technique for improved electric field integral equations. *Microw. Opt. Technol. Lett.* **2013**, *55*, 304–308. [CrossRef]
11. Wu, S.-L.; Li, C.-X. A modified SSOR preconditioning strategy for Helmholtz equations. *J. Appl. Math.* **2012**, *2012*, 365124. [CrossRef]

12. Zhang, L.T.; Cheng, S.H. Modified block symmetric SOR preconditioners for large sparse saddle-point problems. *Appl. Mech. Mater.* **2013**, *241–244*, 2583–2586. [CrossRef]

13. Huang, Y.-M.; Lu, D.-Y. A preconditioned conjugate gradient method for multiplicative half-quadratic image restoration. *Appl. Math. Comput.* **2013**, *219*, 6556–6564. [CrossRef]

14. Meng, Z.; Li, F.; Xu, X.; Huang, D.; Zhang, D. Fast inversion of gravity data using the symmetric successive over-relaxation (SSOR) preconditioned conjugate gradient algorithm. *Explor. Geophys.* **2017**, *48*, 294–304. [CrossRef]

15. Helfenstein, R.; Koko, J. Parallel preconditioned conjugate gradient algorithm on GPU. *J. Comput. Appl. Math.* **2012**, *236*, 3584–3590. [CrossRef]

16. Meyer, K. Technical note: A successive over-relaxation preconditioner to solve mixed model equations for genetic evaluation. *J. Anim. Sci.* **2016**, *94*, 4530–4535. [CrossRef] [PubMed]

17. Sanjuan, G.; Margalef, T.; Cortés, A. Accelerating preconditioned conjugate gradient solver in wind field calculation. In Proceedings of the 2016 International Conference on High Performance Computing & Simulation (HPCS), Innsbruck, Austria, 18–22 July 2016; Smari, W.W., Ed.; IEEE: New York, NY, USA, 2016; pp. 294–301. [CrossRef]

18. Duff, I.S.; Meurant, G.A. The effect of ordering on preconditioned conjugate gradients. *BIT Numer. Math.* **1989**, *29*, 635–657. [CrossRef]

19. DeLong, M.A.; Ortega, J.M. SOR as a preconditioner. *Appl. Numer. Math.* **1995**, *18*, 431–440. [CrossRef]

20. Chen, Y.; Zhao, Y.; Zhao, W.; Zhao, L. A comparative study of preconditioners for GPU-accelerated conjugate gradient solver. In Proceedings of the 2013 IEEE 10th International Conference on High Performance Computing and Communications & 2013 IEEE International Conference on Embedded and Ubiquitous Computing, Zhangjiajie, China, 13–15 November 2013; IEEE: New York, NY, USA, 2013; pp. 628–635. [CrossRef]

21. Li, J.; Zhang, N.-M. A triple-parameter modified SSOR method for solving singular saddle point problems. *BIT Numer. Math.* **2016**, *56*, 501–521. [CrossRef]

22. Pan, C. On generalized SSOR-like iteration method for saddle point problems. *WSEAS Trans. Math.* **2017**, *16*, 239–247.

23. Liang, Z.-Z.; Zhang, G.-F. Modified unsymmetric SOR method for saddle-point problems. *Appl. Math. Comput.* **2014**, *234*, 584–598. [CrossRef]

24. Wang, H.-D.; Huang, Z.-D. On a new SSOR-like method with four parameters for the augmented systems. *East Asian J. Appl. Math.* **2017**, *7*, 82–100. [CrossRef]

25. Louka, M.A.; Missirlis, N.M. A comparison of the extrapolated successive overrelaxation and the preconditioned simultaneous displacement methods for augmented linear systems. *Numer. Math.* **2015**, *131*, 517–540. [CrossRef]

26. Najafi, H.S.; Edalatpanah, S.A. On the modified symmetric successive over-relaxation method for augmented systems. *Comp. Appl. Math.* **2015**, *34*, 607–617. [CrossRef]

27. Darvishi, M.T.; Hessari, P. A modified symmetric successive overrelaxation method for augmented systems. *Comput. Math. Appl.* **2011**, *61*, 3128–3135. [CrossRef]

28. Salkuyeh, D.K.; Shamsi, S.; Sadeghi, A. An improved symmetric SOR iterative method for augmented systems. *Tamkang J. Math.* **2012**, *43*, 479–490. [CrossRef]

29. Huang, Z.-G.; Wang, L.-G.; Xu, Z.; Cui, J.-J. Preconditioned accelerated generalized successive overrelaxation method for solving complex symmetric linear systems. *Comput. Math. Appl.* **2019**, *77*, 1902–1916. [CrossRef]

30. Edalatpour, V.; Hezari, D.; Salkuyeh, D.K. Accelerated generalized SOR method for a class of complex systems of linear equations. *Math. Commun.* **2015**, *20*, 37–52.

31. Hezari, D.; Edalatpour, V.; Salkuyeh, D.K. Preconditioned GSOR iterative method for a class of complex symmetric system of linear equations. *Numer. Linear Algebra Appl.* **2015**, *22*, 761–776. [CrossRef]

32. Salkuyeh, D.K.; Hezari, D.; Edalatpour, V. Generalized successive overrelaxation iterative method for a class of complex symmetric linear system of equations. *Int. J. Comput. Math.* **2015**, *92*, 802–815. [CrossRef]

33. Pu, Z.-N.; Wang, X.-Z. Block preconditioned SSOR methods for H-matrices linear systems. *J. Appl. Math.* **2013**, *2013*, 213659. [CrossRef]

34. Kushida, N. Condition number estimation of preconditioned matrices. *PLoS ONE* **2015**, *10*, e0122331. [CrossRef] [PubMed]

35. Zhang, C.; Miao, G.; Zhu, Y. Convergence on successive over-relaxed iterative methods for non-Hermitian positive definite linear systems. *J. Inequal. Appl.* **2016**, *2016*, 156. [CrossRef]
36. Wang, H.-D.; Huang, Z.-D. On convergence and semi-convergence of SSOR-like methods for augmented linear systems. *Appl. Math. Comput.* **2018**, *326*, 87–104. [CrossRef]
37. Liang, Z.-Z.; Zhang, G.-F. On SSOR iteration method for a class of block two-by-two linear systems. *Numer. Algorithms* **2016**, *71*, 655–671. [CrossRef]
38. Zhou, L.; Zhang, N. Semi-convergence analysis of GMSSOR methods for singular saddle point problems. *Comput. Math. Appl.* **2014**, *68*, 596–605. [CrossRef]
39. Cao, G.; Huang, Y.; Song, Y. Convergence of parallel block SSOR multisplitting method for block H-matrix. *Calcolo* **2013**, *50*, 239–253. [CrossRef]
40. Yang, S.; Gobbert, M.K. The optimal relaxation parameter for the SOR method applied to the Poisson equation in any space dimensions. *Appl. Math. Lett.* **2009**, *22*, 325–331. [CrossRef]
41. Barrett, R.; Berry, M.W.; Chan, T.F.; Demmel, J.; Donato, J.; Dongarra, J.; Eijkhout, V.; Pozo, R.; Romine, C.; Van der Vorst, H. *Templates for the Solution of Linear Systems: Building Blocks for Iterative Methods*, 2nd ed.; Society for Industrial and Applied Mathematics (SIAM): Philadelphia, PA, USA, 1994; pp. 35–49.
42. Turek, V. On improving computational efficiency of simplified fluid flow models. *Chem. Eng. Trans.* **2018**, *70*, 1447–1452. [CrossRef]
43. Van Doormaal, J.P.; Raithby, G.D. Enhancement of the SIMPLE method for predicting incompressible fluid flows. *Numer. Heat Transf.* **1984**, *7*, 147–163. [CrossRef]
44. Patankar, S.V. A calculation procedure for two-dimensional elliptic situations. *Numer. Heat Transf.* **1981**, *4*, 409–425. [CrossRef]
45. Hestenes, M.R.; Stiefel, E. Methods of conjugate gradients for solving linear systems. *J. Res. Natl. Bur. Stand.* **1952**, *49*, 409–436. [CrossRef]
46. Sleijpen, G.L.; Fokkema, D.R. BiCGstab(l) for linear equations involving unsymmetric matrices with complex spectrum. *Electron. Trans. Numer. Anal.* **1993**, *1*, 11–32.

 © 2019 by the author. Licensee MDPI, Basel, Switzerland. This article is an open access article distributed under the terms and conditions of the Creative Commons Attribution (CC BY) license (http://creativecommons.org/licenses/by/4.0/).

Article

One Convenient Method to Calculate Performance and Optimize Configuration for Annular Radiator Using Heat Transfer Unit Simulation

Zhe Xu [1,2,*], Yingqing Guo [1], Huarui Yang [2], Haotian Mao [1], Zongling Yu [2] and Rui Li [2]

[1] School of Power and Energy, Northwestern Polytechnical University, Xi'an 710129, China; yqguo@nwpu.edu.cn (Y.G.); alexmao@mail.nwpu.edu.cn (H.M.)
[2] Xinxiang Aviation Industry (Group) Co., LTD., AVIC, Xinxiang 453049, China; hryangxh@163.com (H.Y.); zonglingy@163.com (Z.Y.); ruiliixh@163.com (R.L.)
*   Correspondence: zhexu@mail.nwpu.edu.cn

Received: 4 December 2019; Accepted: 3 January 2020; Published: 5 January 2020

Abstract: In order to calculate heat transfer capacity and air-side pressure drop of an annular radiator (AR), one performance calculation method was proposed combining heat transfer unit (HTU) simulation and plate-and-fin heat exchanger (PFHX) performance calculation formulas. This method can obtain performance data with no need for meshing AR as a whole, which can be convenient and time-saving, as grid number is reduced in this way. It demonstrates the feasibility of this performance calculation method for engineering applications. In addition, based on the performance calculation method, one configuration optimization method for AR using nondominated sorted genetic algorithm-II (NSGA-II) was also proposed. Fin height (FH) and number of fins in circumferential direction (NFCD) were optimized to maximize heat transfer capacity and minimize air-side pressure drop. Three optimal configurations were obtained from the Pareto optimal points. The heat transfer capacity of the optimal configurations increased by 22.65% on average compared with the original configuration, while the air-side pressure drop decreased by 33.99% on average. It indicates that this configuration optimization method is valid and can provide a significant guidance for AR design.

Keywords: annular radiator; performance calculation; configuration optimization; heat transfer unit; plate-and-fin heat exchanger; nondominated sorted genetic algorithm-II

---

1. Introduction

Heat exchangers, which can transfer heat between two fluids of different temperatures, are widely used in several industrial applications, just like aerospace engineering, petrochemical progress, nuclear power plants, oil refining, and so on [1]. As performance calculation and configuration optimization are main steps in the process of heat exchanger design, numerous up-to-date technologies have been applied in this research area from various points of view.

Theoretical analysis is a common method to calculate heat exchanger performance. As one kind of well-known theoretical analysis method, the Bell–Delaware method is widely used in performance calculation of heat exchangers. In the research of L. Yi et al. [2], one cooling water heat exchanger was designed preliminarily using the Bell–Delaware method, and the design results were verified by HTRI 6.0 commercial software. Some mathematical relationship formulas for performance calculation of shell-and-tube heat exchanger (STHX) with helical baffles were proposed by B. A. Abdelkader et al. [3] based on Bell–Delaware method. In addition, in the research of B. A. Abdelkader et al. [4], Kern method, Bell–Delaware method, and flow stream analysis (Wills Johnston) methods were applied, respectively, to predict both heat-transfer coefficient and pressure drop on the shell side of a heat exchanger.

The calculation results were compared with the experimental data, and the comparison results showed that the Bell–Delaware method was the most accurate method.

Computational fluid dynamics (CFD) technology is also widely used in performance analysis of heat exchangers. The performances of STHXs with and without baffles were compared through CFD calculation program OpenFOAM-2.2.0 by E. Pal et al. [5]. In the research of J. Du et al. [6], a midtemperature gravity heat pipe exchanger was taken as the research object, and the effects of different operating parameters and fin parameters on heat transfer performance were studied using Fluent software. A multitube tank was proposed by M. Ramadan et al. [7] as a heat exchanger. The performance of it was analyzed using CFD simulation code, and the best scenario among three different configurations was obtained by an optimization procedure. Y. Yang et al. [8] studied heat transfer and flow characteristics in a type of plate heat exchanger by numerical simulation, and the correlations of single-phase heat transfer coefficient and friction coefficient were presented. In the research of plate-and-fin heat exchangers (PFHXs), the effects of inlet header configuration on fluid flow maldistribution and the effects of top bypass flow of fins on thermal performance were studied by A. Raul et al. [9] and H. Cai et al. [10] using CFD simulation, respectively. In addition, based on CFD technology, the performances of STHXs separately with segmental baffles [11,12], trefoil-hole baffles [12,13], and helical baffles [11,12,14] were analyzed. Especially in the research of A. El Maakoul et al. [12], the simulation results of these three kinds of heat exchangers were compared, and STHX with helical baffles was found to be the one that had the best comprehensive performance.

Combining theoretical analysis with CFD simulation can be another useful way to research heat exchanger performance. R. Amini et al. [15] compared the CFD simulation results with the calculation results by the Bell–Delaware method to validate the accuracy of the simulation method. D.M. Godino et al. [16] calculated the heat transfer coefficient of the preheater separately by Bell–Delaware method and Kern method, and the calculation results were both compared with the CFD simulation data to analyze the accuracy. X. Gu et al. [17] came up with periodic whole cross-section computation models to obtain performance data of segmental baffle heat exchanger, shutter baffle heat exchanger, and trapezoid-like tilted baffle heat exchanger, and the reliability of the method was verified by comparing the simulation data with the calculation results using the Bell–Delaware method. I. Milcheva et al. [18] improved the traditional Bell–Delaware method and introduced an enhancement factor to calculate the performance of a STHX with double-segmental baffles. The calculation results were compared with the CFD simulation data to validate the effectiveness of this method.

As for configuration optimization for heat exchanger, nondominated sorted genetic algorithm-II (NSGA-II) is commonly used in some research. The Bell–Delaware procedure and the ε-NTU method were applied in STHX performance estimation by S. Sanaye et al. [19] and NSGA-II was used to maximize heat transfer coefficient and minimize pressure drop simultaneously. M. Chahartaghi et al. [20] combined NSGA-II with entransy dissipation theory to minimize the entransy dissipation numbers separately caused by thermal conduction and fluid friction for STHX. Z. Xu et al. [21] calculated the performances of two kinds of STHXs by theoretical formulas and optimized their structural parameters using NSGA-II, respectively. In order to optimize the configuration of a PFHX with offset strip fins, NSGA-II combined with the ε-NTU method was adopted in the research of R. Song et al. [22]. In addition, as a popular optimization method, NSGA-II combined with response surface method was also used in the configuration optimizations of STHX with helical baffles [23–25], spiral-wound heat exchanger [26], helically coiled tube heat exchanger [27], torsional flow heat exchanger [28], and triple concentric-tube heat exchanger [29]. Except for NSGA-II, some other novel optimization algorithms were also proposed to optimize configuration parameters of different heat exchangers, such as firefly algorithm [30], Tsallis differential evolution algorithm [31], bat algorithm [32], Taguchi method [33], particle swarm optimization (PSO) [34], cohort intelligence algorithm [35], tree traversal method [36], surrogate-based optimization algorithm [37], wale optimization [38], topology optimization [39], Jaya algorithm [40], and so on.

As annular radiator (AR) is a neoteric heat exchanger, the research about its thermal-hydraulic performance calculation and configuration optimization, which are badly in need of design processes, are rare. In this paper, a feasible and reliable method for performance calculation and configuration optimization based on heat transfer unit (HTU) simulation and NSGA-II was proposed, which can conveniently obtain AR heat transfer capacity and air-side pressure drop, while avoiding the problem of the huge amount of grids generated by meshing AR as a whole directly, and getting the optimized fin height (FH) and number of fins in circumferential direction (NFCD), which are a trade-off on maximizing heat transfer capacity and minimizing air-side pressure drop.

2. Method

2.1. Physical Model and Normal Operating Conditions

AR is an air–liquid heat exchanger, which mainly consists of liquid channels, annular substrate, and fins. As depicted in Figure 1, liquid flows through the channels inside the annular substrate, and air passes by the fins, which are averagely distributed on the circular inner side of the annular substrate. So, the heat can be transferred from the liquid of higher temperature to the air of lower temperature through fins and annular substrates.

The material of AR is aluminum alloy, and the normal operating conditions are shown in Table 1. In this case, the working fluid of liquid is lubricating oil.

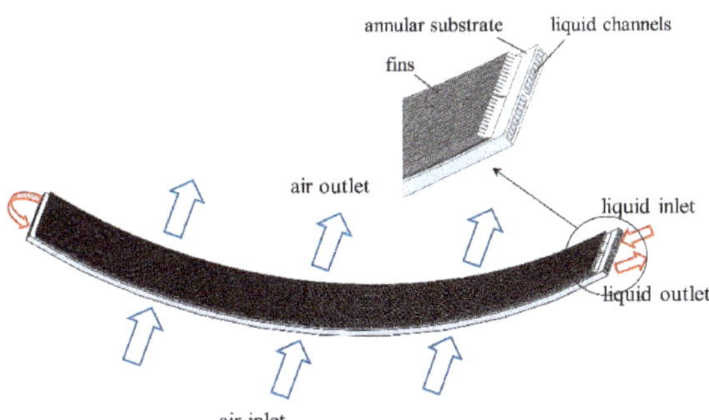

Figure 1. Schematic diagram of annular radiator (AR).

Table 1. Normal operating conditions of AR.

Normal Operating Condition	Value
Liquid inlet volume flow rate (L/min)	30
Liquid inlet temperature (K)	423
Air inlet mass flow rate (kg/h)	5500
Air inlet temperature (K)	286
Air inlet pressure (kPa)	5.1

2.2. Performance Calculation Method

2.2.1. Structural Equivalence

As AR is structurally similar to PFHX with fins of rectangular straight wave, the performance calculation formulas of this kind of heat exchanger were applied to the calculation of AR heat transfer capacity in this method. Basic formulas of PFHX heat transfer capacity calculation are given in Appendix A. The structure of the equivalent PFHX is depicted in Figure 2. The main configuration parameters of AR and the corresponding PFHX are shown in Figure 3 and Table 2.

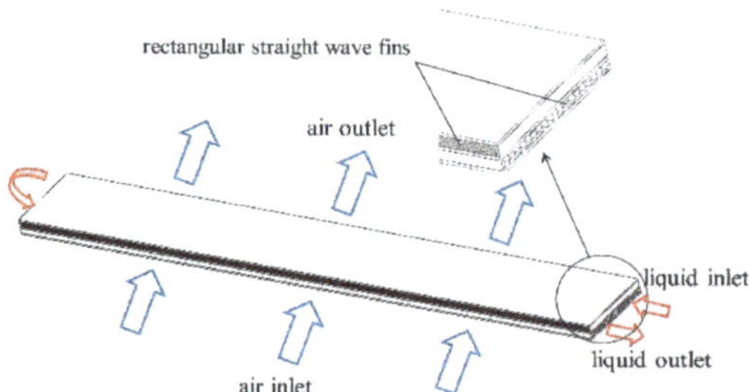

Figure 2. Structure of the equivalent plate-and-fin heat exchanger (PFHX).

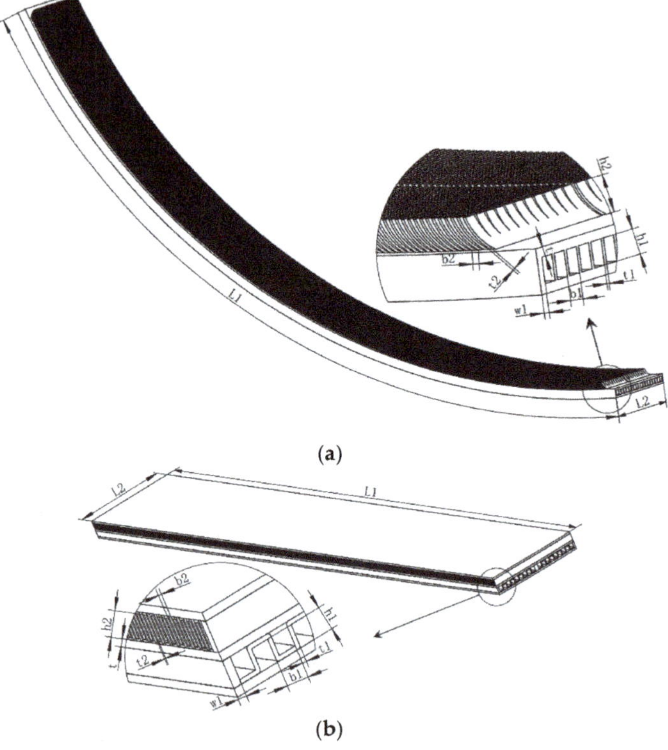

(**a**)

(**b**)

Figure 3. Configuration parameter correspondence between AR and PFHX: (**a**) configuration parameters of AR; (**b**) corresponding configuration parameters of PFHX.

Table 2. Configuration parameter values.

Configuration Parameter	Value (mm)	Configuration Parameter	Value (mm)
L1	1185	L2	150
b1	7.8	b2	2.1
h1	8	h2	10
t1	1.5	t2	0.8
w1	3	t	3

2.2.2. HTU Simulation Method

A traditional way to obtain the performance data of a heat exchanger is through emulating it as a whole directly. While, for AR simulation, a huge number of grids will be generated by this traditional method in the process of meshing. That is extremely time-consuming for solving and is beyond the available computing resources. In order to avoid this problem, HTU simulation, rather than overall simulation of AR, was used in this method. When an AR is working, the air above the fins of the AR does not pass through the clearance between fins, which means that it is unscientific if the air inlet flow rate of AR is used in the PFHX performance calculation formulas directly. So, the percentage of the air calculated in the formulas needs to be obtained through HTU simulation in this method. This percentage is defined by Equation (1):

$$k = q_{fin} / q_{HTU}. \tag{1}$$

In Equation (1), q_{fin} and q_{HTU} are the volume flow rates of the air passing through the clearance between fins of HTU and the air of HTU air-side, respectively, and k reflects the percentage of the air that can be calculated in PFHX performance calculation formulas.

HTUs are obtained through dividing AR, and two kinds of HTUs can be obtained according to the symmetry of AR, as shown in Figure 4. As HTU A and HTU B are structurally similar, and air flows through HTU A first, HTU A was chosen to be the only one kind of HTU to be emulated in this method. In this case, the central angle of HTU was 2 degrees.

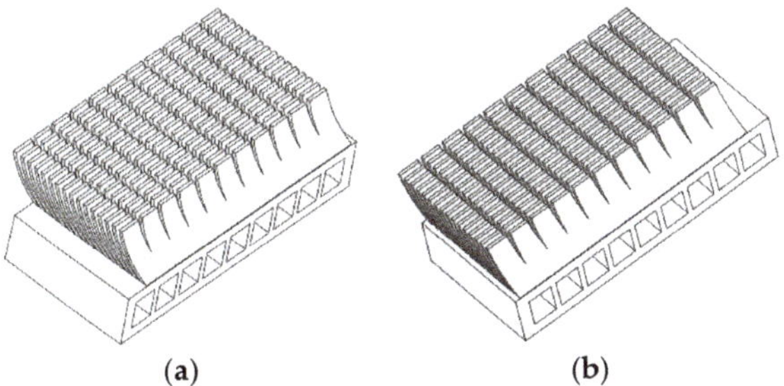

(a)   (b)

Figure 4. Two kinds of heat transfer units (HTUs): (**a**) HTU A; (**b**) HTU B.

In the process of CFD simulation, ANSYS commercial software was adopted. There were two fluid domains, including air domain and liquid domain, and one solid domain in the CFD simulation of HTU, as shown in Figure 5. Unstructured grids were adopted, and the normal operating conditions of AR were used in HTU simulation. In order to improve simulation accuracy, grid independence validation was carried out. In Figure 6, $\Delta P'$ represents air-side pressure drop of HTU. As shown in this figure, the simulation deviations of k and $\Delta P'$ were both in the acceptable range when the grid number increases above the point of 24,020,699. So, the mesh generation settings of this point were used.

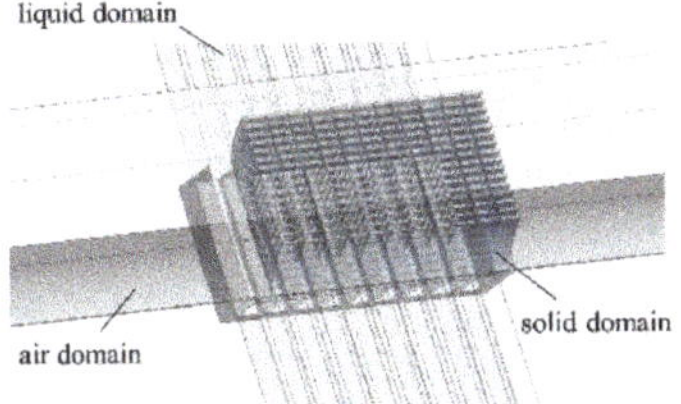

Figure 5. HTU computational domains.

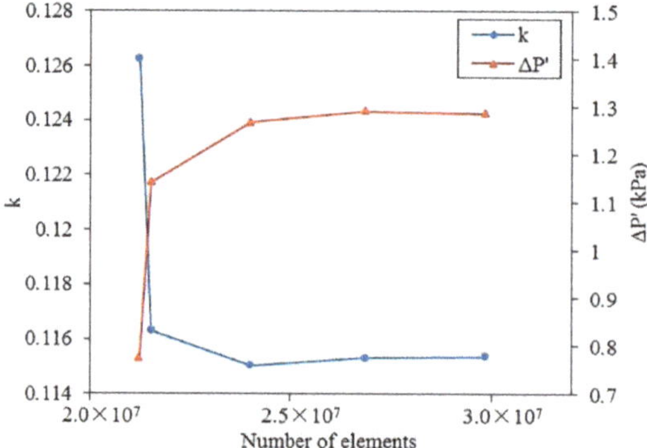

Figure 6. Grid independence validation.

In this case, the air flow rate was less than Ma0.3, and the temperature deviation between air inlet and outlet was small. So, like liquid, air was treated as a Newtonian and incompressible fluid with constant physical property parameters. In addition, in the process of simulation, fluid flow and heat transfer process were considered as turbulent and in steady-state, and fouling resistance was neglected.

In this method, a realizable k-ε turbulence model and the default constant values were adopted. Velocity-inlet and pressure-outlet were separately used as inlets and outlets of both air and liquid. In the process of solving this kind of fluid flow heat transfer problem, the SIMPLE (semi-implicit method for pressure-linked equations) algorithm is the most widely used. The core of the algorithm is to use continuous equations and momentum equations to construct an approximate pressure correction equation on staggered grids to calculate pressure field and correct velocity. The SIMPLE algorithm is very useful in engineering applications. However, the convergence velocity of the SIMPLE algorithm is not fast. Compared with the SIMPLE algorithm, the SIMPLEC (SIMPLE-consistent) algorithm, which is the improved version of SIMPLE algorithm, can achieve a higher rate of convergence as it synchronizes speed field improvement process with pressure field improvement process [41]. Thus, in this study, SIMPLEC algorithm, as one built-in solution algorithm in ANSYS Fluent commercial software, was chosen to solve the simulation problem. The second order upwind was used for the momentum, turbulent kinetic energy, turbulent dissipation rate, and energy. The default convergence criterion was adopted, which is that the normalized residuals are less than 1×10^{-6} for energy equation and 1×10^{-3} for the other equations. Iteration number was set as 2000.

2.2.3. Method Procedure

AR performance calculation in this method included heat transfer capacity calculation and air-side pressure drop calculation. The main procedures are shown as follows:

1. Input the configuration parameters of AR as shown in Figure 3 and Table 2, and input the normal operating conditions of AR shown in Table 1.
2. Obtain HTU through dividing AR, and emulate HTU under the normal operating conditions of AR.
3. According to HTU simulation results, acquire k and $\Delta P'$.
4. Calculate the air inlet volume flow rate of AR, which can be used in PFHX performance calculation formulas by Equation (2):

$$q_v = kq_v'$$

(2)

In Equation (2), q_v is the air inlet volume flow rate of AR used in PFHX performance calculation formulas and q_v' is the air inlet volume flow rate of AR.

5. Calculate air-side pressure drop of AR. As HTU is obtained by cutting AR along its symmetry plane, air-side pressure drop of AR can be computed by Equation (3):

$$\Delta P = 2\Delta P'$$ (3)

In Equation (3), ΔP is air-side pressure drop of AR.

6. Calculate heat transfer capacity of AR using PFHX performance calculation formulas.

   (a) Assume one heat transfer capacity for AR.
   (b) Calculate the mean temperature of liquid/air-side under the assumptive heat transfer capacity.
   (c) Calculate the physical property parameters (such as fluid density, kinematic viscosity, etc.) of liquid/air-side under the mean temperature of the corresponding side.
   (d) Calculate the effective heat transfer area values and the heat transfer coefficients of liquid-side and air-side.
   (e) Calculate NTU (number of transfer units), heat transfer efficiency, and heat transfer capacity.
   (f) Give the value of the calculated heat transfer capacity to the assumptive heat transfer capacity, and go to step (b) until the deviation of them is within acceptable limits.
   (g) Output the calculated heat transfer capacity.

The flowchart of this method is shown in Figure 7.

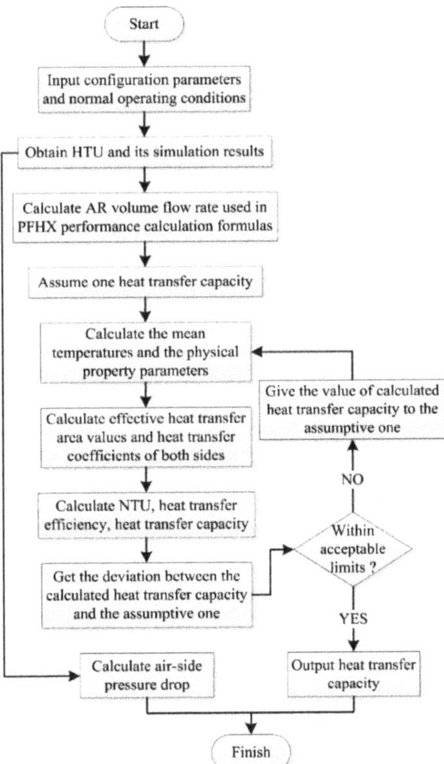

Figure 7. Flowchart of AR performance calculation method.

2.3. Configuration Optimization Method

In this research, in consideration of the performances of AR, heat transfer capacity Q and air-side pressure drop ΔP were chosen as two objective functions, and FH and NFCD were regarded as two design parameters. Thus, the multiobjective optimization problem was be formulated as Equation (4):

$$\begin{cases} Min & -Q, \Delta P \\ S.t. & 7\text{mm} \leq FH \leq 16\text{mm} \& FH \in Z \\ & 238 \leq NFCD \leq 544 \& NFCD \in Z \end{cases} . \tag{4}$$

For a multiobjective optimization problem, it is always hard to find a solution that is absolutely optimal. However, through some evolutional algorithms, a Pareto optimal front that contains a series of optimal solutions can be obtained. NSGA-II is one kind of evolutional algorithm that is based on a genetic algorithm for multiobjective optimization. One or more satisfactory solutions can be selected by some criteria from the Pareto optimal front obtained by NSGA-II [24]. It incorporates elitism [42], and no sharing parameter needs to be chosen a priori [43]. As the nondominated sorting method is used as the ranking scheme in this algorithm, the convergence velocity of NSGA-II is faster than the traditional Pareto ranking method. Besides, as the constraint handling method also uses a nondominance principle as the objective, which guarantees that the feasible solutions are always ranked higher than the unfeasible solutions, penalty functions and Lagrange multipliers are not needed in NSGA-II [44]. So, due to these advantages, and in order to optimize these two configuration parameters shown in Equation (4) under the normal operating conditions shown in Table 1, the multiobjective genetic algorithm NSGA-II is adopted in this configuration optimization method to obtain the optimal solutions. The main procedures are shown as follows:

1. The two design parameters and the constraints of them, and the two conflicting optimization objectives are initialized.
2. Emulate HTUs of some different values of the two design parameters and record the corresponding simulation data of k and $\Delta P'$.
3. Obtain the functional relationships shown in Equations (5) and (6) by fitting using the simulation data calculated in step 2.

$$k = f_1(FH, NFCD), \tag{5}$$

$$\Delta P' = f_2(FH, NFCD) \tag{6}$$

4. Optimize design parameters using NSGA-II based on the AR performance calculation method in this research and the functional relationships obtained in step 3.

   (a) Initialize NSGA-II parameters, including population size, generation number, and so on, and generate a random population in the constraints of design parameters.
   (b) Calculate objective function values for each chromosome of the population using AR performance calculation method and the functional relationships obtained in step 3.
   (c) Sort chromosomes based on non-domination and crowding distance. In this method, the crowding distance is compared only if the ranks for both chromosomes are the same.
   (d) Choose the chromosomes that are fit for reproduction as the parents of the next generation using tournament algorithm.
   (e) Generate children by crossover and mutation, and calculate their objective function values using AR performance calculation method and the functional relationships obtained in step 3.
   (f) Combine parents and children, and sort them based on nondomination and crowding distance.
   (g) A new generation is extracted based on ranking.

(h)   The above procedures are repeated from step (d) until convergence.

(i)   Output the Pareto optimal front which consists of a series of solutions.

Other details of NSGA-II are referred in [43]. The flowchart of this method is shown in Figure 8.

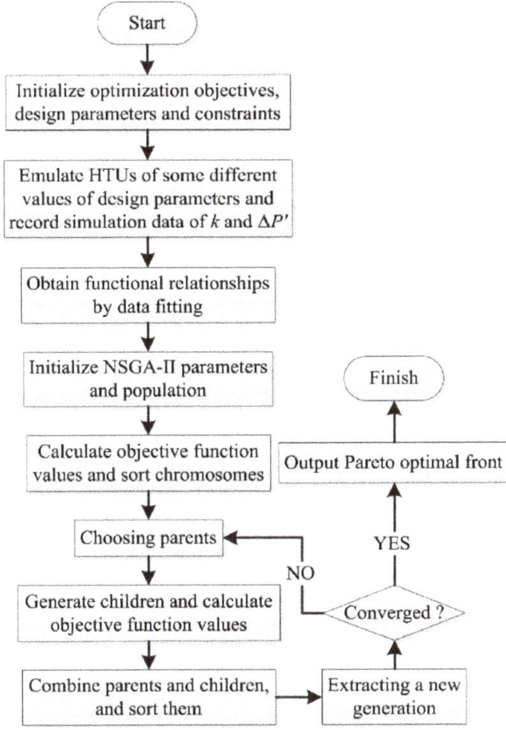

Figure 8. Flowchart of AR configuration optimization method.

3. Results and Discussion

3.1. Performance Calculation Method Validation

HTU simulation results are shown in Figure 9. As depicted, when air passes by fins, only a small part of air passes through the clearance between fins, and the velocity of the air goes above fins is higher than that of the other. So, the calculation of k is significant. Based on the plane perpendicular to the air flow direction and located in the middle place of the finned area, the calculated k was 11.5%. That means only 11.5% of the air passes through the clearance between fins and can be used in PFHX performance calculation formulas.

In order to validate the accuracy of the AR performance calculation method, the calculation results were compared with the experimental data. The experimental conditions, the experimental schematic diagram, and the testing equipment for AR are shown in Table 1 and Figure 10; Figure 11, respectively. The experimental equipment included an air supply system, oil circulation system, and measuring system. An air supply system can provide AR the airflow of the required flow rate, temperature, and pressure. The maximum air supply capacity of it is 8600 kg/h, which can meet the requirement of this test. An oil circulation system can heat the lubricating oil to the required temperature and then supply it to AR. The measuring system includes some instruments distributed in the inlets and outlets of both air and oil. As the radius of AR is large, five pressure sensors and five temperature sensors were averagely distributed in the air inlet of AR, and the same is true for the air outlet. The experimental temperature or pressure value of the air inlet or outlet of AR was computed by averaging the corresponding five measured values. The results of the comparison between calculation results and experimental data are given in Figure 12.

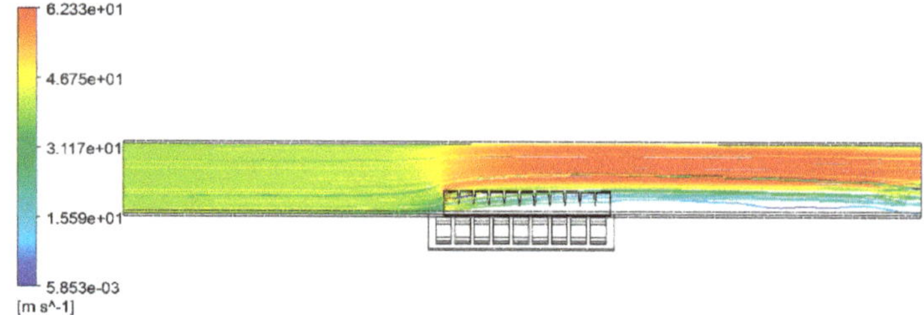

Figure 9. Velocity streamline of HTU air-side.

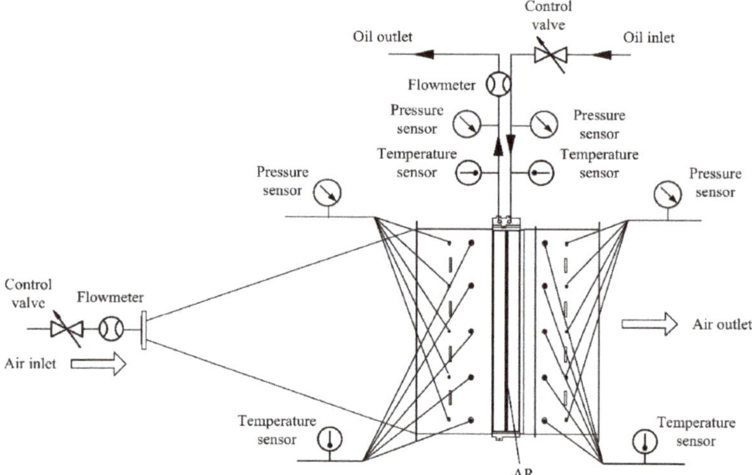

Figure 10. Experimental schematic diagram.

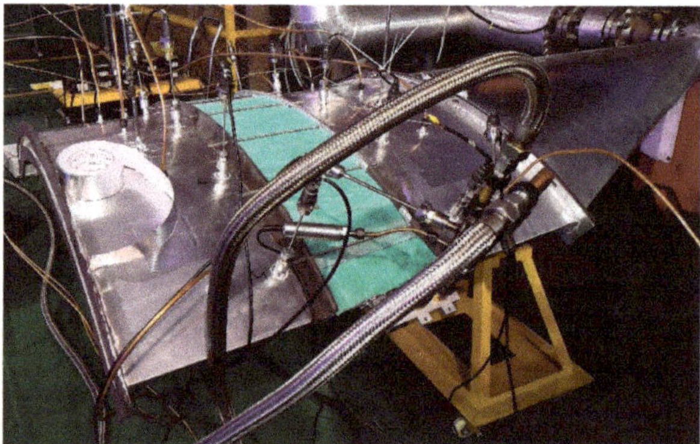

Figure 11. Testing equipment for AR.

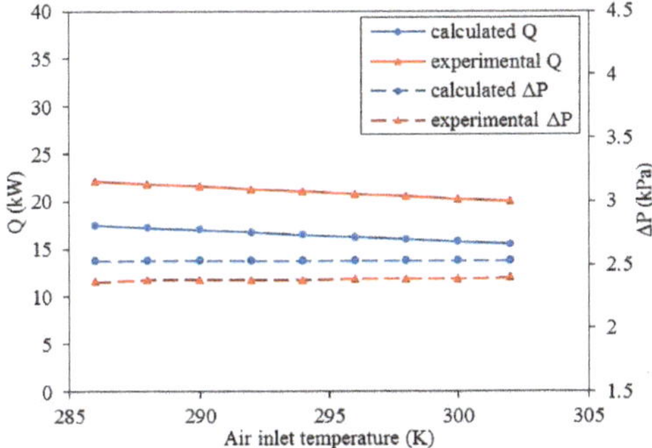

Figure 12. Comparison between calculation results and experimental data.

As depicted in Figure 12, ΔP obtained by this method was consistent with the experimental data, while the error of Q was higher than ΔP. The deviations of Q and ΔP were 20.7–22.4% and 5.5–6.9% with the average errors of 21.5% and 6.2%, respectively. The reason for this is that, as shown in Equation (3), half of ΔP is directly obtained by CFD simulation, which is accurate. However, the calculation of Q using PFHX performance calculation formulas only can consider the heat transfer of the air that flows through the clearances of fins, while the air that flows above fins also participates in the heat transfer process. Even though the heat transfer effect of the air flowing above fins is very small, it can still influence the precision of Q calculation and make the calculated values by this method lower than the experimental data. Thus, if a highly accurate Q result is needed, this part of air also needs to be considered, which is not realizable through using PFHX theoretical formulae directly. AR needs to be emulated as a whole directly to meet the high accuracy requirement, which is too time-consuming to realize in engineering applications. In the design process of engineering applications, the precision of this method is enough for AR performance prediction, and it is more convenient and time saving. Hence, it can be concluded that this method is feasible for engineering applications, and can be used in the configuration optimization process in this research.

Ignoring factors of this method itself, there are also some other reasons that can cause the differences between the calculation results and the experimental data, just like deviations of formulas themselves, simplification of simulation model, and unavoidable experimental errors.

3.2. Design Parameter Effects and Configuration Optimization Results

3.2.1. Functional Relationships Fitting

In this case, 16 sets of HTU simulation data under different values of FH and NFCD were averagely obtained within design parameter constraints, and the surfaces of fitting were obtained using least square method and polynomial fitting based on these data. The fitting surfaces are shown in Figures 13 and 14, and the functional relationships can be represented by Equations (7) and (8).

$$k = 0.03211 + 0.03009x - 0.0001297y + 0.0002819x^2 - 3.732 \times 10^{-5}xy$$
$$+4.445 \times 10^{-8}y^2 - 8.684 \times 10^{-6}x^3 + 1.589 \times 10^{-6}x^2y - 2.16 \times 10^{-8}xy^2 \tag{7}$$

$$\Delta P = -0.6039 + 0.1434x + 0.004469y - 0.009944x^2 - 0.000316xy$$
$$-1.103 \times 10^{-5}y^2 + 0.000389x^3 - 8.93 \times 10^{-6}x^2y + 1.378 \times 10^{-6}xy^2 \tag{8}$$

Sum of squares due to error (SSE) and the coefficient of determination R^2 are significant goodness of fit criteria to evaluate the accuracy of fitting function [45]. The closer the SSE and R^2 are to 0 and 1, respectively, the better is the fitting function. The SSEs of Equations (7) and (8) are 1.9739×10^{-5} and

0.0033, and the R^2 values of that are 0.9999 and 0.9993, respectively. Hence, the functional relationships can be obtained accurately.

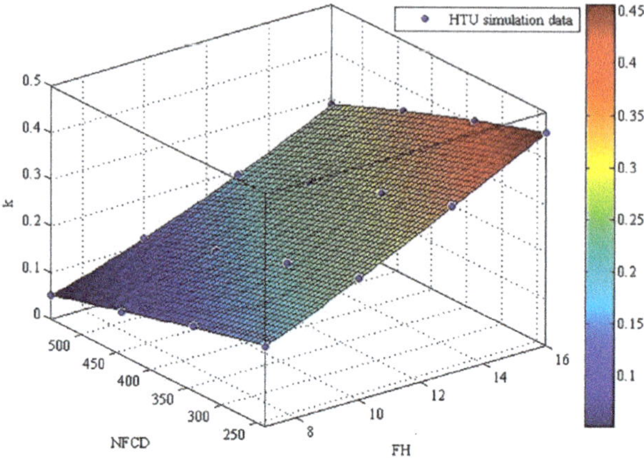

Figure 13. Fitting surface for the percentage of the air that can be calculated in PFHX performance calculation formulas (k).

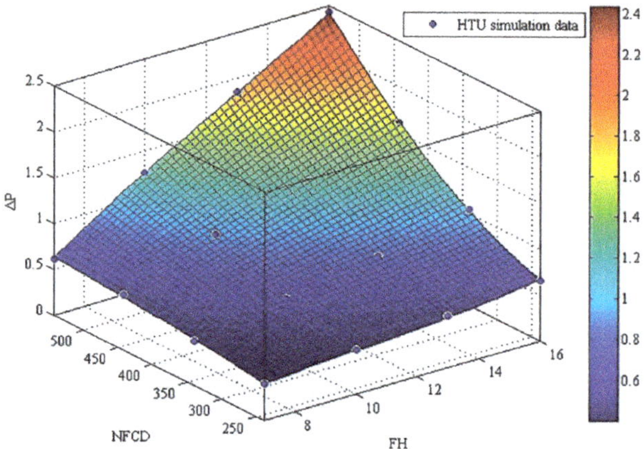

Figure 14. Fitting surface for air-side pressure drop of AR (ΔP).

3.2.2. Design Parameter Effects

The effects of design parameters on Q and ΔP were analyzed based on Equations (7) and (8) and the AR performance calculation method. The effects of FH are represented in Figure 15, while NFCD was 544. As depicted, while FH increased from 7 mm to 16 mm, Q and ΔP increased by 9.66–26.26 kW and 1.25–4.87 kPa, respectively. The growth of FH led to the increase of heat transfer area, which can lead to the increase of Q. At the same time, as FH rose, the influence area of fins rose, and the resistance influence of fins was enhanced. Thus, ΔP increased continuously.

The effects of NFCD are shown in Figure 16, while FH was 10 mm. As depicted, while NFCD increased from 238 to 544, ΔP increased by 0.96–2.53 kPa, and Q increased from 16.71 kW first, reaching its highest point of 19.77 kW when NFCD was 408, and then decreased to 17.25 kW. The increase of NFCD indicates the growth of the heat transfer area, which results in the increase of Q at first. Meanwhile the increase of NFCD can also lead to the decrease of the clearance space between two adjacent fins, which means that air is increasingly harder to flow through these clearances and the velocity of air that flows through the air channel without fins increases. So, ΔP increased continuously in this process, and k decreased continuously. The decrease of k will weaken the heat exchange capability of AR. After NFCD increases to a certain level, the decrease of k can lead to the decline of Q, even though heat transfer area is still increasing. Thus, the curve of Q rose first and then fell after the point.

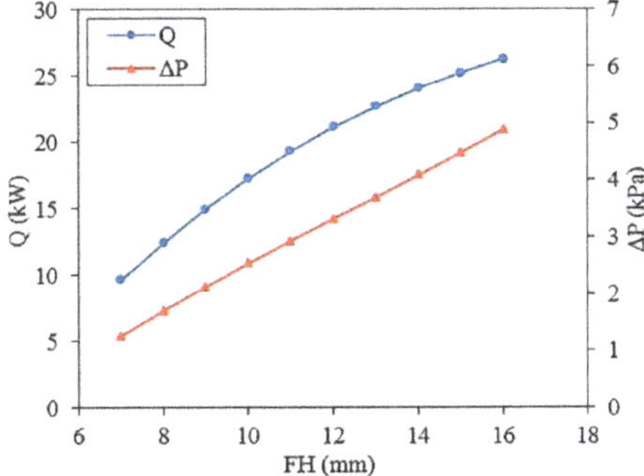

Figure 15. Heat transfer capacity (Q) and ΔP separately versus fin height (FH).

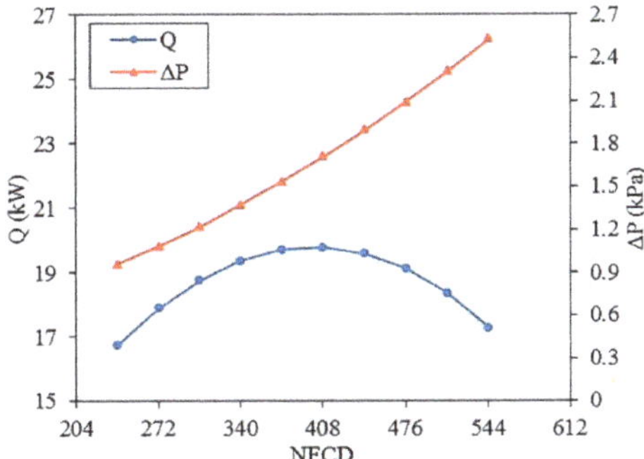

Figure 16. Q and ΔP separately versus number of fins in circumferential direction (NFCD).

3.2.3. Optimization Results

In order to consider the comprehensive performance of AR, the multiobjective configuration optimization method driven by NSGA-II was conducted. The conflicting optimization objectives were set as the minimization of $-Q$ and ΔP both. Population size, maximum iteration number, analog binary cross distribution index, and polynomial mutation distribution index were set as 100, 500, 20, and 20, respectively. The obtained Pareto optimal points are shown in Figure 17.

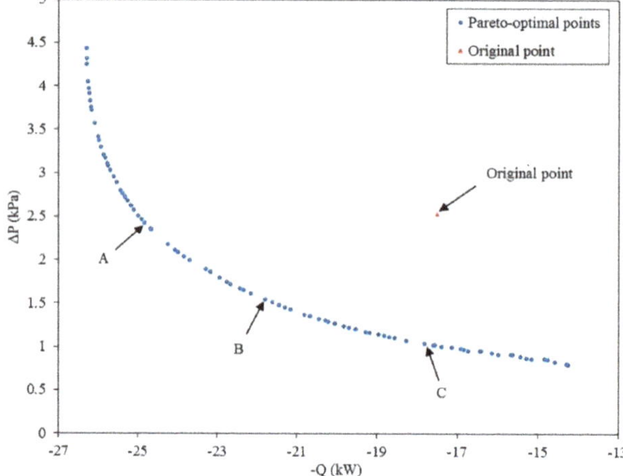

Figure 17. Pareto optimal points.

As depicted in Figure 17, both the values of $-Q$ and ΔP of some obtained optimal points were lower than that of the original point, which means that these obtained configurations had better comprehensive performances than the original one. In addition, three Pareto optimal solutions were chosen and are shown in Figure 17 and Table 3. The selection method of these three Pareto optimal solutions follows the following principles:

1. $-Q$ of the point A is much lower than the original point, while ΔP of it is only a little less than the original point.
2. $-Q$ of the point C is only a little lower than the original point, while ΔP of it is much less than the original point.
3. The point B is chosen from the points around the middle position in the range from the point A to the point C.

Table 3. Optimization performances comparison of AR.

Parameters	FH (mm)	NFCD	Q (kW)	ΔP (kPa)
Original point	10	544	17.50	2.53
Optimal point A	16	375	24.81	2.42
Optimal point B	15	290	21.79	1.55
Optimal point C	11	248	17.79	1.04

Compared with the original point shown in Table 3, Q values of the point A, B, and C separately increased by 41.77%, 24.51%, and 1.66%, and ΔP values of them decreased by 4.35%, 38.74%, and 58.89%, respectively. It is clearly shown that Q of the optimal configurations increased by 22.65% on average, while ΔP decreased by 33.99% on average. Based on the above comparisons, it indicates that the proposed configuration optimization method is valid and feasible, and the comprehensive performance of AR can be enhanced by this method.

4. Conclusions

In this research, a performance calculation method for AR was proposed and verified by experiment. Heat transfer capacity and air-side pressure drop were calculated through combining HTU simulation and PFHX performance calculation formulas, rather than through emulating AR as a whole directly. So, the problem of the huge amount of grids generated by meshing AR can be effectively avoided, which means that this method is convenient to use and can save calculation time and computing resources. It demonstrates the feasibility of this performance calculation method for engineering applications.

Based on this performance calculation method, a configuration optimization method for AR was also come up with using NSGA-II in this research. Heat transfer capacity maximization and air-side pressure drop minimization were regarded as two conflicting objectives, and FH and NFCD were set as two design parameters. A set of Pareto optimal solutions were obtained and some of them had better comprehensive performances than the original configurations. Three optimal solutions were chosen and compared with the original configuration. The comparison results illustrate that the heat transfer capacity of the optimal configurations increased by 22.65% on average compared with the original configuration, while the air-side pressure drop decreased by 33.99% on average. It indicates that this configuration optimization method is valid and can provide a significant guidance for AR design.

Author Contributions: Conceptualization, Z.X.; investigation, Z.X. and H.M.; methodology, Z.X.; project administration, Y.G. and H.Y.; resources, Z.X. and Z.Y.; software, Z.X.; supervision, Y.G. and H.Y.; validation, Z.X.; visualization, Z.X.; writing—original draft, Z.X.; writing—review and editing, Z.X., Y.G., Z.Y., and R.L. All authors have read and agreed to the published version of the manuscript.

Funding: This research received no external funding.

Conflicts of Interest: The authors declare no conflict of interest.

Nomenclature

Latin letters

A	heat transfer area	m^2
C^*	heat capacity rate ratio	-
c_p	specific heat capacity	J/(kg·K)
h	wave height of fins	m
NTU	number of transfer units	-
Pr	Prandtl number	-
Q	heat transfer capacity	W
q_m	mass flow rate	kg/s
q_v	volume flow rate	L/min
R_w	wall thermal resistance	K/W
T	inlet temperature	K
U	total heat transfer coefficient	$W/(m^2·K)$
W	thermal capacity rate	W/K
Z	set of integer	-

Greek letters

α	heat transfer coefficient	$W/(m^2·K)$
δ	thickness	m
ΔP	pressure drop	Pa
η	heat transfer efficiency	-
η_A	surface efficiency of air-side	-
η_{fA}	fin efficiency of air-side	-
η_{fL}	fin efficiency of liquid-side	-
η_L	surface efficiency of liquid-side	-
ω	mass flow rate per square meter	$kg/(m^2·s)$
λ	heat conductivity	W/(m·K)

Subscripts

A	air
f	fins
L	liquid
p	division plate
w	wall

Appendix A

The basic formula of heat transfer capacity calculation of PFHX is shown by Equation (A1) [46]:

$$Q = \eta W_{\min}(T_L - T_A). \tag{A1}$$

As the numbers of air passes and liquid passes are 1 and 2, respectively, η is given by Equations (A2)–(A7) [47]:

$$\eta = \frac{\left(\frac{1-C^*\eta_i}{1-\eta_i}\right)^2 - 1}{\left(\frac{1-C^*\eta_i}{1-\eta_i}\right)^2 - C^*}, \tag{A2}$$

$$\eta_i = 1 - \exp\left\{\frac{NTU^{0.22}}{C^*}[\exp(-C^*NTU^{0.78}) - 1]\right\} \tag{A3}$$

$$NTU = \frac{UA}{W_{\min}} \tag{A4}$$

$$C^* = \frac{W_{\min}}{W_{\max}} \tag{A5}$$

$$\frac{1}{UA} = \frac{1}{\eta_L \alpha_L A_L} + R_w + \frac{1}{\eta_A \alpha_A A_A} \tag{A6}$$

$$R_w = \frac{\delta_p}{\lambda_w A_p} \tag{A7}$$

In Equation (A6), η_L and η_A reflect surface efficiencies of liquid-side and air-side, respectively, and A_L and A_A separately represent heat transfer areas of liquid-side and air-side. These variables can be calculated by Equations (A8)–(A15) [47,48]:

$$\eta_L = 1 - \frac{A_{fL}}{A_L}(1 - \eta_{fL}), \tag{A8}$$

$$\eta_A = 1 - \frac{A_{fA}}{A_A}(1 - \eta_{fA}) \tag{A9}$$

$$A_L = A_p + A_{fL} \tag{A10}$$

$$A_A = A_p + A_{fA} \tag{A11}$$

$$\eta_{fL} = \frac{\tanh(m_L h_L)}{m_L h_L} \tag{A12}$$

$$\eta_{fA} = \frac{\tanh(m_A h_A)}{m_A h_A} \tag{A13}$$

$$m_L = \sqrt{\frac{2\alpha_L}{\lambda_{fL}\delta_{fL}}} \tag{A14}$$

$$m_A = \sqrt{\frac{2\alpha_A}{\lambda_{fA}\delta_{fA}}} \tag{A15}$$

W and α refer to thermal capacity rate and heat transfer coefficient, respectively, and they are given by Equations (A16) and (A17) [46,48]:

$$W = q_m c_p \tag{A16}$$

$$\alpha = \frac{j\omega c_p}{Pr^{2/3}} \tag{A17}$$

References

1. Bayram, H.; Sevilgen, G. Numerical investigation of the effect of variable baffle spacing on the thermal performance of a Shell and tube heat Exchanger. *Energies* **2017**, *10*, 1156. [CrossRef]
2. Yi, L.; Lei, C.; Jinrong, Q. One Method to Design a Heat Exchanger for Equipment Cooling Water. In *25th International Conference on Nuclear Engineering*; American Society of Mechanical Engineers Digital Collection: Shanghai, China, 2017.
3. Abdelkader, B.A.; Jamil, M.A.; Zubair, S.M. Thermal-hydraulic characteristics of helical baffle shell-and-tube heat exchangers. *Heat Transf. Eng.* **2019**, 1–34. [CrossRef]
4. Abdelkader, B.A.; Zubair, S.M. The Effect of a Number of Baffles on the performance of Shell-and-Tube Heat Exchangers. *Heat Transf. Eng.* **2019**, *40*, 39–52. [CrossRef]
5. Pal, E.; Kumar, I.; Joshi, J.B.; Maheshwari, N. CFD simulations of shell-side flow in a shell-and-tube type heat exchanger with and without baffles. *Chem. Eng. Sci.* **2016**, *143*, 314–340. [CrossRef]
6. Du, J.; Wu, X.; Li, R.; Cheng, R. Numerical Simulation and Optimization of MID-Temperature Heat Pipe Exchanger. *Fluid Dyn. Mater. Process.* **2019**, *15*, 77–87. [CrossRef]
7. Ramadan, M.; Khaled, M.; Jaber, H.; Faraj, J.; Bazzi, H.; Lemenand, T. Numerical simulation of Multi-Tube Tank heat exchanger: Optimization analysis. *Energy Sour. Part A Recovery Util. Environ. Eff.* **2019**, 1–11. [CrossRef]
8. Yang, Y.; Hu, H.; Su, P.; Mei, X.; Sun, W. Characteristic Analysis and Structure Optimization of a Type of Plate Heat Exchanger. In *IOP Conference Series: Earth and Environmental Science*; IOP Publishing: Bristol, UK, 2019; p. 042055.
9. Raul, A.; Bhasme, B.; Maurya, R. A numerical investigation of fluid flow maldistribution in inlet header configuration of plate fin heat exchanger. *Energy Procedia* **2016**, *90*, 267–275. [CrossRef]
10. Cai, H.; Su, L.; Liao, Y.; Weng, Z. Numerical and experimental study on the influence of top bypass flow on the performance of plate fin heat exchanger. *Appl. Therm. Eng.* **2019**, *146*, 356–363. [CrossRef]
11. Shinde, S.; Hadgekar, P.; Pavithran, S. Comparative thermal analysis of helixchanger with segmental heat exchanger using bell-delaware method. *Int. J. Adv. Eng. Technol.* **2012**, *3*, 235.
12. El Maakoul, A.; Laknizi, A.; Saadeddine, S.; El Metoui, M.; Zaite, A.; Meziane, M.; Abdellah, A.B. Numerical comparison of shell-side performance for shell and tube heat exchangers with trefoil-hole, helical and segmental baffles. *Appl. Therm. Eng.* **2016**, *109*, 175–185. [CrossRef]
13. Wang, K.; Bai, C.; Wang, Y.; Liu, M. Flow dead zone analysis and structure optimization for the trefoil-baffle heat exchanger. *Int. J. Therm. Sci.* **2019**, *140*, 127–134. [CrossRef]
14. Wang, S.; Xiao, J.; Ye, S.; Song, C.; Wen, J. Numerical investigation on pre-heating of coal water slurry in shell-and-tube heat exchangers with fold helical baffles. *Int. J. Heat Mass Transf.* **2018**, *126*, 1347–1355. [CrossRef]
15. Amini, R.; Amini, M.; Jafarinia, A.; Kashfi, M. Numerical investigation on effects of using segmented and helical tube fins on thermal performance and efficiency of a shell and tube heat exchanger. *Appl. Therm. Eng.* **2018**, *138*, 750–760. [CrossRef]
16. Godino, D.M.; Corzo, S.F.; Nigro, N.M.; Ramajo, D.E. CFD simulation of the pre-heater of a nuclear facility steam generator using a thermal coupled model. *Nucl. Eng. Des.* **2018**, *335*, 265–278. [CrossRef]
17. Gu, X.; Zheng, Z.; Xiong, X.; Wang, T.; Luo, Y.; Wang, K. Characteristics of Fluid Flow and Heat Transfer in the Shell Side of the Trapezoidal-like Tilted Baffles Heat Exchanger. *J. Therm. Sci.* **2018**, *27*, 602–610. [CrossRef]
18. Milcheva, I.; Heberle, F.; Brüggemann, D. Modeling and simulation of a shell-and-tube heat exchanger for Organic Rankine Cycle systems with double-segmental baffles by adapting the Bell-Delaware method. *Appl. Therm. Eng.* **2017**, *126*, 507–517. [CrossRef]
19. Sanaye, S.; Hajabdollahi, H. Multi-objective optimization of shell and tube heat exchangers. *Appl. Therm. Eng.* **2010**, *30*, 1937–1945. [CrossRef]
20. Chahartaghi, M.; Eslami, P.; Naminezhad, A. Effectiveness improvement and optimization of shell-and-tube heat exchanger with entransy method. *Heat Mass Transf.* **2018**, *54*, 3771–3784. [CrossRef]
21. Xu, Z.; Guo, Y.; Mao, H.; Yang, F. Configuration Optimization and Performance Comparison of STHX-DDB and STHX-SB by A Multi-Objective Genetic Algorithm. *Energies* **2019**, *12*, 1794. [CrossRef]
22. Song, R.; Cui, M. Single-and multi-objective optimization of a plate-fin heat exchanger with offset strip fins adopting the genetic algorithm. *Appl. Therm. Eng.* **2019**, *159*, 113881. [CrossRef]

23. Wen, J.; Gu, X.; Wang, M.; Wang, S.; Tu, J. Numerical investigation on the multi-objective optimization of a shell-and-tube heat exchanger with helical baffles. *Int. Commun. Heat Mass Transf.* **2017**, *89*, 91–97. [CrossRef]

24. Wang, S.; Xiao, J.; Wang, J.; Jian, G.; Wen, J.; Zhang, Z. Configuration optimization of shell-and-tube heat exchangers with helical baffles using multi-objective genetic algorithm based on fluid-structure interaction. *Int. Commun. Heat Mass Transf.* **2017**, *85*, 62–69. [CrossRef]

25. Wang, S.; Xiao, J.; Wang, J.; Jian, G.; Wen, J.; Zhang, Z. Application of response surface method and multi-objective genetic algorithm to configuration optimization of Shell-and-tube heat exchanger with fold helical baffles. *Appl. Therm. Eng.* **2018**, *129*, 512–520. [CrossRef]

26. Wang, S.; Jian, G.; Xiao, J.; Wen, J.; Zhang, Z.; Tu, J. Fluid-thermal-structural analysis and structural optimization of spiral-wound heat exchanger. *Int. Commun. Heat Mass Transf.* **2018**, *95*, 42–52. [CrossRef]

27. Wang, G.; Wang, D.; Deng, J.; Lyu, Y.; Pei, Y.; Xiang, S. Experimental and numerical study on the heat transfer and flow characteristics in shell side of helically coiled tube heat exchanger based on multi-objective optimization. *Int. J. Heat Mass Transf.* **2019**, *137*, 349–364. [CrossRef]

28. Gu, X.; Wang, T.; Chen, W.; Luo, Y.; Tao, Z. Multi-objective Optimization on Structural Parameters of Torsional Flow Heat Exchanger. *Appl. Therm. Eng.* **2019**, *161*, 113831. [CrossRef]

29. Amanuel, T.; Mishra, M. Thermohydraulic optimization of triple concentric-tube heat exchanger: A multi-objective approach. *Proc. Inst. Mech. Eng. Part E J. Process Mech. Eng.* **2019**, *233*, 589–600. [CrossRef]

30. Mohanty, D.K. Application of firefly algorithm for design optimization of a shell and tube heat exchanger from economic point of view. *Int. J. Therm. Sci.* **2016**, *102*, 228–238. [CrossRef]

31. De Vasconcelos Segundo, E.H.; Amoroso, A.L.; Mariani, V.C.; Dos Santos Coelho, L. Economic optimization design for shell-and-tube heat exchangers by a Tsallis differential evolution. *Appl. Therm. Eng.* **2017**, *111*, 143–151. [CrossRef]

32. Tharakeshwar, T.; Seetharamu, K.; Prasad, B.D. Multi-objective optimization using bat algorithm for shell and tube heat exchangers. *Appl. Therm. Eng.* **2017**, *110*, 1029–1038. [CrossRef]

33. Etghani, M.M.; Baboli, S.A.H. Numerical investigation and optimization of heat transfer and exergy loss in shell and helical tube heat exchanger. *Appl. Therm. Eng.* **2017**, *121*, 294–301. [CrossRef]

34. Esfe, M.H.; Mahian, O.; Hajmohammad, M.H.; Wongwises, S. Design of a heat exchanger working with organic nanofluids using multi-objective particle swarm optimization algorithm and response surface method. *Int. J. Heat Mass Transf.* **2018**, *119*, 922–930. [CrossRef]

35. Dhavle, S.V.; Kulkarni, A.J.; Shastri, A.; Kale, I.R. Design and economic optimization of shell-and-tube heat exchanger using cohort intelligence algorithm. *Neural Comput. Appl.* **2018**, *30*, 111–125. [CrossRef]

36. Hao, J.-H.; Chen, Q.; Ren, J.-X.; Zhang, M.-Q.; Ai, J. An experimental study on the offset-strip fin geometry optimization of a plate-fin heat exchanger based on the heat current model. *Appl. Therm. Eng.* **2019**, *154*, 111–119. [CrossRef]

37. Chien, N.B.; Linh, N.X.; Jong-Taek, O. Numerical Optimization of Flow Distribution inside Inlet Header of Heat Exchanger. *Energy Procedia* **2019**, *158*, 5488–5493. [CrossRef]

38. Kumar, S.D.; Chandramohan, D.; Purushothaman, K.; Sathish, T. Optimal hydraulic and thermal constrain for plate heat exchanger using multi objective wale optimization. *Mater. Today: Proc.* **2019**. [CrossRef]

39. Kobayashi, H.; Yaji, K.; Yamasaki, S.; Fujita, K. Freeform winglet design of fin-and-tube heat exchangers guided by topology optimization. *Appl. Therm. Eng.* **2019**, *161*, 114020. [CrossRef]

40. More, K.C.; Rao, R.V. Design Optimization of Plate-Fin Heat Exchanger by Using Modified Jaya Algorithm. In *Advanced Engineering Optimization through Intelligent Techniques*; Springer: Berlin/Heidelberg, Germany, 2020; pp. 165–172.

41. Ansys Inc. *ANSYS Fluent User's Guide*; ANSYS: Canonsburg, PA, USA, 2018.

42. Yue, S.; Wang, Y.; Wang, H. Design and optimization of tandem arranged cascade in a transonic compressor. *J. Therm. Sci.* **2018**, *27*, 349–358. [CrossRef]

43. Deb, K.; Pratap, A.; Agarwal, S.; Meyarivan, T. A fast and elitist multiobjective genetic algorithm: NSGA-II. *IEEE Trans. Evol. Comput.* **2002**, *6*, 182–197. [CrossRef]

44. Wen, J.; Yang, H.; Tong, X.; Li, K.; Wang, S.; Li, Y. Optimization investigation on configuration parameters of serrated fin in plate-fin heat exchanger using genetic algorithm. *Int. J. Therm. Sci.* **2016**, *101*, 116–125. [CrossRef]

45. Renaud, O.; Victoria-Feser, M.-P. A robust coefficient of determination for regression. *J. Stat. Plan. Inference* **2010**, *140*, 1852–1862. [CrossRef]
46. Shah, R.K.; Sekulic, D.P. *Fundamentals of Heat Exchanger Design*; John Wiley & Sons: Hoboken, NJ, USA, 2003.
47. Yu, J. *Heat Exchanger Principle and Design*; Beijing University of Aeronautics and Astronautics Press: Beijing, China, 2006.
48. Qi, M. *Refrigeration Accessories*; Aviation Industry Press: Beijing, China, 1992.

© 2020 by the authors. Licensee MDPI, Basel, Switzerland. This article is an open access article distributed under the terms and conditions of the Creative Commons Attribution (CC BY) license (http://creativecommons.org/licenses/by/4.0/).

MDPI

St. Alban-Anlage 66

4052 Basel

Switzerland

Tel. +41 61 683 77 34

Fax +41 61 302 89 18

www.mdpi.com

Energies Editorial Office

E-mail: energies@mdpi.com

www.mdpi.com/journal/energies

www.ingramcontent.com/pod-product-compliance
Lightning Source LLC
LaVergne TN
LVHW070619170726
843515LV00004B/128